Multi-Stage Actuation Systems and Control

Multi-Stage Actuation Systems and Control

Chunling Du and Chee Khiang Pang

CRC Press is an imprint of the
Taylor & Francis Group, an **informa** business

CRC Press
Taylor & Francis Group
6000 Broken Sound Parkway NW, Suite 300
Boca Raton, FL 33487-2742

© 2019 by Taylor & Francis Group, LLC
CRC Press is an imprint of Taylor & Francis Group, an Informa business

No claim to original U.S. Government works

Printed on acid-free paper

International Standard Book Number-13: 978-1-138-48075-9 (Hardback)

This book contains information obtained from authentic and highly regarded sources. Reasonable efforts have been made to publish reliable data and information, but the author and publisher cannot assume responsibility for the validity of all materials or the consequences of their use. The authors and publishers have attempted to trace the copyright holders of all material reproduced in this publication and apologize to copyright holders if permission to publish in this form has not been obtained. If any copyright material has not been acknowledged, please write and let us know so we may rectify in any future reprint.

Except as permitted under U.S. Copyright Law, no part of this book may be reprinted, reproduced, transmitted, or utilized in any form by any electronic, mechanical, or other means, now known or hereafter invented, including photocopying, microfilming, and recording, or in any information storage or retrieval system, without written permission from the publishers.

For permission to photocopy or use material electronically from this work, please access www. copyright.com (http://www.copyright.com/) or contact the Copyright Clearance Center, Inc. (CCC), 222 Rosewood Drive, Danvers, MA 01923, 978-750-8400. CCC is a not-for-profit organization that provides licenses and registration for a variety of users. For organizations that have been granted a photocopy license by the CCC, a separate system of payment has been arranged.

Trademark Notice: Product or corporate names may be trademarks or registered trademarks and are used only for identification and explanation without intent to infringe.

Library of Congress Cataloging-in-Publication Data

Names: Du, Chunling, author. | Pang, Chee Khiang, author.
Title: Multi-stage actuation systems and control / Chunling Du and Chee Khiang Pang.
Description: First edition. | Boca Raton, FL : CRC Press/Taylor & Francis Group, 2018. | Includes bibliographical references.
Identifiers: LCCN 2018032541 | ISBN 9781138480759 (hardback : acid-free paper) | ISBN 9781351062183 (ebook)
Subjects: LCSH: Microactuators. | Automatic control.
Classification: LCC TJ223.A25 D83 2018 | DDC 629.8—dc23
LC record available at https://lccn.loc.gov/2018032541

Visit the Taylor & Francis Web site at
http://www.taylorandfrancis.com

and the CRC Press Web site at
http://www.crcpress.com

To my families.

C. Du

To tbt. For our adventure.

C. K. Pang

Contents

Preface..xi
Authors..xiii
List of Abbreviations xv

1. Mechanical Actuation Systems..1
    1.1   Introduction 1
    1.2   Brief Description of Application Examples.......................................2
          1.2.1   Dual-Stage Actuation Systems...2
          1.2.2   Micro X–Y Stages..3
    1.3   Actuators 4
          1.3.1   Primary Actuator..4
          1.3.2   Secondary Actuators 6
    1.4   Single-Stage Actuation Systems..8
    1.5   Dual-Stage Actuation Systems 9
    1.6   Three-Stage Actuation Systems 10
    References 11

2. High-Precision Positioning Control of Dual-Stage Actuation Systems...15
    2.1   Introduction 15
    2.2   Control Schemes...15
    2.3   Controller Design Method in the Continuous-Time Domain 18
    2.4   Controller Design Method in the Discrete-Time Domain 22
    2.5   Application in the Controller Design of a Dual-Stage Actuation System 25
          2.5.1   Actuator Modeling Based on Frequency Responses Measurement..25
          2.5.2   Controller Design and Simulation 25
          2.5.3   Experimental Results 29
    2.6   Conclusion...29
    References 31

3. Control of Thermal Microactuator-Based, Dual-Stage Actuation Systems 33
    3.1   Introduction 33
    3.2   Modeling of a Thermal Microactuator..33
    3.3   Controller Design and Performance Evaluation 37
    3.4   Experimental Results...41
    3.5   Conclusion...45
    References 45

vii

viii   *Contents*

4. Modeling and Control of a Three-Stage Actuation System 47
  4.1 Introduction 47
  4.2 Actuator and Vibration Modeling 47
  4.3 Control Strategy and Controller Design 52
  4.4 Performance Evaluation 56
  4.5 Experimental Results 60
  4.6 Different Configurations of the Control System 61
  4.7 Conclusion 66
  References 67

**5. Dual-Stage System Control Considering Secondary Actuator
   Stroke Limitation** 69
  5.1 Introduction 69
  5.2 More Freedom Loop Shaping for Microactuator
       Controller Design 70
  5.3 Dual-Stage System Control Design for 5 kHz Bandwidth 70
  5.4 Evaluation with the Consideration of External Vibration
       and Microactuator Stroke 74
  5.5 Conclusion 76
  References 76

**6. Saturation Control for Microactuators in Dual-Stage
   Actuation Systems** 79
  6.1 Introduction 79
  6.2 Modeling and Feedback Control 79
  6.3 Anti-Windup Compensation Design 80
  6.4 Simulation and Experimental Results 82
  6.5 Conclusion 87
  References 87

**7. Time Delay and Sampling Rate Effect on Control Performance
   of Dual-Stage Actuation Systems** 89
  7.1 Introduction 89
  7.2 Modeling of Time Delay 90
  7.3 Dual-Stage Actuation System Modeling with Time Delay for
       Controller Design 91
  7.4 Controller Design with Time Delay for the Dual-Stage
       Actuation Systems 93
  7.5 Time Delay Effect on Dual-Stage System Control Performance 95
       7.5.1 $(\tau_m + \tau_c) < T_s$ 95
       7.5.2 $T_s < (\tau_m + \tau_c) < 2T_s$ 97
  7.6 Sampling Rate Effect on Dual-Stage System Control
       Performance 99
  7.7 Conclusion 103
  References 104

Contents

ix

8. PZT Hysteresis Modeling and Compensation 105
    8.1   Introduction 105
    8.2   Modeling of Hysteresis 105
        8.2.1   PI Model 106
        8.2.2   GPI Model 107
        8.2.3   Inverse GPI Model 108
    8.3   Application of GPI Model to a PZT-Actuated Structure 109
        8.3.1   Modeling of the Hysteresis in the PZT-Actuated
               Structure.. 110
        8.3.2   Hysteresis Compensator Design................................... 112
        8.3.3   Experimental Verification... 112
    8.4   Conclusion... 116
    References 117

9. Seeking Control of Dual-Stage Actuation Systems with
Trajectory Optimization.. 119
    9.1   Introduction 119
    9.2   Current Profile of VCM Primary Actuator................................ 120
        9.2.1   PTOS Method 120
        9.2.2   A General Form of VCM Current Profiles.................... 121
    9.3   Control System Structure for the Dual-Stage Actuation System....121
    9.4   Design of VCM Current Profile a_v and Dual-Stage Reference
        Trajectory r_d 123
    9.5   Seeking within PZT Milliactuator Stroke 127
    9.6   Seeking over PZT Milliactuator Stroke...................................... 128
    9.7   Conclusion... 129
    References 131

10. High-Frequency Vibration Control Using PZT Active Damping 133
    10.1   Introduction 133
    10.2   Singular Perturbation Method-Based Controller Design 134
        10.2.1 Singular Perturbation Control Topology......................... 134
        10.2.2 Identification of Fast Dynamics Using PZT as a Sensor.... 135
        10.2.3 Design of Controllers 136
             10.2.3.1 Fast Subsystem Estimator $\tilde{G}_V^*$ 137
             10.2.3.2 Fast Controller $\tilde{C}_V$ 137
             10.2.3.3 Slow Controller $\bar{C}_V$ 137
        10.2.4 Simulation and Experimental Results 138
             10.2.4.1 Frequency Responses 138
             10.2.4.2 Time Responses 139
    10.3 H_2 Controller Design 141
    10.4   Design of $C_d(z)$ with H_2 Method and Notch Filters 143
    10.5   Design of Mixed H_2/H_∞ Controller $C_d(z)$............................... 144
    10.6   Application Results.. 146
        10.6.1 System Modeling 146

10.6.2 H_2 Active Damping Control	147
10.6.3 Mixed H_2/H_∞ Active Damping Control	149
10.6.4 Experimental Results	150
10.7 Conclusion	150
References	152

11. Self-Sensing Actuation of Dual-Stage Systems ... 155
11.1 Introduction	155
11.2 Estimation of PZT Secondary Actuator's Displacement y_P^*	156
11.2.1 Self-Sensing Actuation and Bridge Circuit	156
11.2.2 PZT Displacement Estimation Circuit H_B	157
11.3 Design of Controllers	160
11.3.1 VCM Controller and Controller C_D	160
11.3.2 PZT Controller	161
11.4 Performance Evaluation	163
11.4.1 Effectiveness of C_D	163
11.4.2 Position Errors	163
11.5 Conclusion	166
References	166

12. Modeling and Control of a MEMS Micro X–Y Stage Media Platform ... 167
12.1 Introduction	167
12.2 MEMS Micro X–Y Stage	168
12.2.1 Design and Simulation of Micro X–Y Stage	168
12.2.1.1 Static	169
12.2.1.2 Dynamic	170
12.2.2 Modeling of Micro X–Y Stage	171
12.2.3 Fabrication of the MEMS Micro X–Y Stage	173
12.3 Capacitive Self-Sensing Actuation	176
12.3.1 Design of CSSA Bridge Circuit	176
12.3.2 Experimental Verification	178
12.4 Robust Decoupling Controller Design	179
12.4.1 Choice of Pre-Shaping Filters	179
12.4.2 Controller Synthesis	180
12.4.3 Frequency Responses	182
12.4.4 Time Responses	182
12.4.5 Robustness Analysis	185
12.5 Conclusion	186
References	186

13. Conclusions ... 189

Index ... 191

Preface

The demands for the applications of miniature, micro-, and even nanoscale devices are growing, leading to an interest in the development of micro-systems capable of carrying out fine tasks with a miniature, micro-, and even nanoscale performance, thus substantially extending the range of applications of traditional motion systems. Conventional motion concepts are no longer able to fulfill the demands concerning miniaturization, high accuracy, and reliability. Use of microactuators as fine actuation components of dual-stage or multi-stage systems allows the motion to be realized with fast response and micrometer or even nanometer accuracy, and high reliability.

Microactuators and their systems have significant problems in regard to design concepts, control principles, accuracy, resistance to environmental influences, etc. For effective applications, the problems in control principle are regarded as challenges for research and industry as the problems are specific to systems with different microactuators and must be approached from different points of view in order to effectively make use of the microactuators.

This book is designed for academic researchers and engineering practitioners in systems and control, especially those engaged in the area of control in mechanical systems with microactuators and multi-stage actuations. The book aims at empowering readers with a clear understanding of multi-stage mechanism, different microactuators' performances, their limitations to control system performance, and problems encountered in control system design and techniques for solving these problems and dealing with these limitations. Special attention is given to recently developed control techniques in three-stage systems and MEMS positioning stage. Real-world examples are given to demonstrate the control techniques such as dual-stage and three-stage systems in hard disk drives and micro X–Y stage in manufacturing systems.

Although the control problem of microactuation systems has been studied for a long time, its work in multi-stage actuation systems remains interesting and indeed becomes more challenging in many applications such as precision engineering and hard disk drives, where an extremely high positioning accuracy and fast response are required. Therefore, the study of control systems for multi-stage actuation systems is significantly necessary for researchers and engineering practitioners. It is our intention in this book to present to readers some of the recent developments in this field. The book focuses on multi-stage system control techniques for high precision positioning and fast response, and demonstration of the benefits gained from the applications of these techniques. A number of simulation and experimental results with comprehensive evaluations are provided in each chapter, except Chapter 1, which is dedicated to the dual-stage and three-stage structures and the review of several typical types of microactuators.

Chunling Du

Chee Khiang Pang

Authors

Dr. Chunling Du earned her B.Eng. and the M.Eng. degrees in electrical engineering from Nanjing University of Science and Technology (NJUST), China, and PhD in electrical engineering from Nanyang Technological University (NTU), Singapore. She has been working on data storage servo control systems for more than 15 years, and has been recently working on advanced and intelligent manufacturing as a senior research and development scientist. She has been interested and working in the areas of data analytics and machine learning for condition monitoring, precision motion control, advanced control algorithms and applications, and real-time sensing and signal processing. She has authored and coauthored more than 100 papers and two books titled H_∞ *Control and Filtering of Two-Dimensional Systems* (Springer, 2002) and *Modeling and Control of Vibration in Mechanical Systems* (CRC Press, 2010). She is a senior member of IEEE.

Dr. Chee Khiang Pang (Justin) earned his B.Eng.(Hons.), M.Eng., and PhD degrees in 2001, 2003, and 2007, respectively, all in electrical and computer engineering, from National University of Singapore (NUS). In 2003, he was a Visiting Fellow in the School of Information Technology and Electrical Engineering (ITEE), University of Queensland (UQ), St Lucia, QLD, Australia. From 2006 to 2008, he was a Researcher (Tenure) with Central Research Laboratory, Hitachi Ltd, Kokubunji, Tokyo, Japan. In 2007, he was a Visiting Academic in the School of ITEE, UQ, St Lucia, QLD, Australia. From 2008 to 2009, he was a Visiting Research Professor in the Automation & Robotics Research Institute (ARRI), University of Texas at Arlington (UTA), Fort Worth, TX, USA. From 2009 to 2016, he was an Assistant Professor in Department of Electrical and Computer Engineering (ECE), NUS, Singapore. Currently, he is an Assistant Professor (Tenure) in Engineering Cluster, Singapore Institute of Technology, Singapore. He is an Adjunct Assistant Professor of NUS, Senior Member of IEEE, and a Member of ASME. He is also the Chairman of IEEE Control Systems Chapter of Singapore. His research interests are in data-driven control and optimization, with realistic applications to robotics, mechatronics, and manufacturing systems. Dr. Pang is an author/editor of four research monographs including *Intelligent Diagnosis and Prognosis of Industrial Networked Systems* (CRC Press, 2011), *High-Speed Precision Motion Control* (CRC Press, 2011), *Advances in High-Performance Motion Control of Mechatronic Systems* (CRC Press, 2013), and *Multi-Stage Actuation Systems and Control* (CRC Press, 2019). He is currently serving as an Executive Editor for *Transactions of the Institute of Measurement and Control*, an Associate Editor for *Asian Journal of Control, IEEE Control Systems Letters, Journal of Defense*

Modeling & Simulation, and *Unmanned Systems*, and is on the Conference Editorial Board for IEEE Control Systems Society (CSS). In recent years, he has also served as a Guest Editor for *Asian Journal of Control*, *International Journal of Automation and Logistics*, *International Journal of Systems Science*, *Journal of Control Theory and Applications*, and *Transactions of the Institute of Measurement and Control*. He was the recipient of The Best Application Paper Award in the 8th Asian Control Conference (ASCC 2011), Kaohsiung, Taiwan, 2011, and the Best Paper Award in the IASTED International Conference on Engineering and Applied Science (EAS 2012), Colombo, Sri Lanka, 2012.

List of Abbreviations

AREs   Algebraic Riccati Equations
BOE   Buffered Oxide Etching
CSSA   Capacitive Self-Sensing Actuation
DRIE   Deep Reactive Ion Etching
FEA   Finite Element Analysis
FFT   Fast Fourier Transform
GPI   Generalized Prandtl-Ishlinskii (PI)
IMC   Internal Model Control
LDV   Laser Doppler Vibrometer
LMI   Linear Matrix Inequality
LPF   Low-Pass Filter
MISO   Multi-Input-Single-Output
MIMO   Multi-Input-Multi-Output
MEMS   Microelectromechanical Systems
MTA   Micro Thermal Actuator
PBSS   Probe-Based Storage Systems
PES   Position Error Signal
DISO   Dual-Input-Single-Output
PID   Proportional-Integral-Derivative
PI   Prandtl-Ishlinskii
PMMA   Polymethyl Methacrylate
PPF   Positive Position Feedback
PTOS   Proximate-Time-Optimal-Servomechanism
PZT   Lead Zirconate Titanate/Pb-Zr-Ti/Piezoelectric
RPM   Rotations Per Minute
SEM   Scanning Electron Microscopy
SSA   Self-Sensing Actuation
VCM   Voice Coil Motor
ZOH   Zero-Order-Hold

1

Mechanical Actuation Systems

1.1 Introduction

Single-stage systems involving one coarse actuator are not powerful enough any longer in the scenarios where high positioning accuracy and high servo bandwidth are demanded, particularly when long stroke motion is required simultaneously. Usually, dual-stage systems composed of a coarse or primary actuator and a fine actuator working together are employed to meet the aforementioned requirements.

Most of the dual-stage systems are applied in hard disk drives [1,2], and at the same time, they have been extended to micro- and nanopositioning applications such as scanning microscopy [3], biological imaging [4], microgripping [5], and nanoassembly [6]. In these applications, many coarse actuators such as DC (direct current) motor, voice coil motor (VCM), permanent magnet stepper motor, and permanent magnet linear synchronous motor are popular selections, and most of these applications use piezoelectric actuator as fine actuator. For example, in [7], a coarse-fine dual-stage nanopositioning system is developed where a permanent magnet stepper motor as a coarse stage and a piezoelectric stack actuator as a fine stage are stacked together; in [8], a dual-stage system is presented where a permanent magnet linear motor is used to drive a coarse stage and a VCM to drive a fine stage; and in [9], the design of a dual-stage actuation system is reported where a linear motor and a piezoelectric actuator are employed as coarse and fine drivers, respectively. For hard disk drives, dual-stage actuation systems are mostly comprised of VCM coarse actuator and PZT (Pb-Zr-Ti) fine actuator assembled at different locations such as suspension, slider, and head. In the application of scanning probe microscopy, to quickly acquire a large specimen's surface information, precise nanopositioning stage with both a long scanning range and a high bandwidth is needed to implement an accurate and rapid scanning. Thus, dual-stage actuation systems have been employed in the nanopositioning stage for the scanning task.

In order to overcome fine actuators' stringent stroke limitation and increase control bandwidth as well, three-stage actuation systems are necessary in practical applications and will be introduced in the later part of this chapter.

1.2 Brief Description of Application Examples

1.2.1 Dual-Stage Actuation Systems

There are five major physical parts in a hard disk drive as seen in Figure 1.1: baseplate and cover, spindle and motor assembly, actuator assembly, disks, head/suspension assembly, and electronics card. Briefly introducing, the spindle and motor assembly includes disk clamps to clamp the disks. In the head/suspension assembly, an air-bearing surface is created on the surface next to the rotating disk, the slider carrying heads flies on top of the disk surface, and a gimbal attaches the slider to the suspension. The electronics card involves the drivers for the spindle motor and actuators, read/write (R/W) electronics, a servo demodulator, and microprocessors for servo control and the control of interface to host computer. The actuator assembly contains an actuator driven by VCM and mounted via ball-bearing at each end of a pivot shaft, a flex cable carrying the heads and actuator leads, arms to support the suspension/head extension between the disks, as well as a secondary actuator driven jointly with the VCM actuator for dual-stage actuation.

Positioning information or servo information is embedded in each disk surface and used to position the magnetic heads on the disk surfaces. Position measurement of the magnetic heads is achieved by means of analyzing the position error signal (PES) calculated from read back signals, i.e., through the process of PES demodulation.

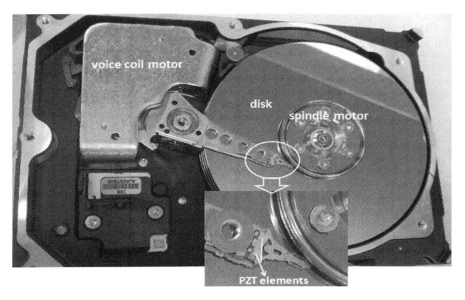

FIGURE 1.1
An internal view of a hard disk drive with PZT milliactuator.

1.2.2 Micro X–Y Stages

Many universities and research institutes around the world have been working on probe-based storage systems (PBSS). While their approaches may differ, the main components used for high-speed parallel read/write/erase (R/W/E) data are similar. In this section, the developed PBSS in the so-called "nanodrive" is shown in Figure 1.2.

In a typical PBSS, the major components include but are not limited to the following: (i) probes (consisting of a sharp tip on a cantilever), (ii) polymer storage medium, (iii) microelectromechanical systems (MEMS) micro X–Y stage or MEMS scanner platform, and (iv) control, signal processing, and sensor electronics, etc.

The control electronics generally consist of a digital signal processor for signal processing, including PES demodulation, read channel encoding/decoding, multiplexing/demultiplexing, control signal computations, etc. Various controllers can take charge of input/output (I/O) scheduling, data distribution and reconstruction, and host interface and failure management. On receiving reference commands, the micro X–Y stage is actuated to the desired locations with the help of thermal [11] or capacitive sensors [12]. Using the written-in information on the dedicated servo fields (where

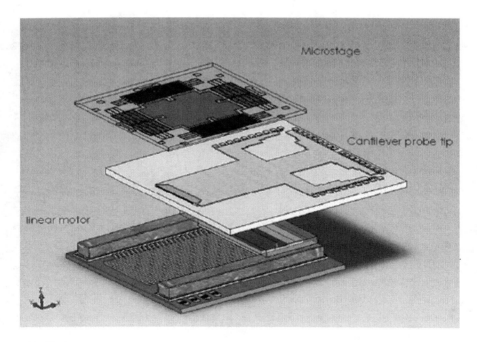

FIGURE 1.2
Components of the developed PBSS in the so-called "nanodrive" consisting of cantilevers carrying probe tips, linear motor, and MEMS X–Y stage with recording medium. (From Ref. [10].)

partitioned areas of the recording medium are specially reserved for servo data where only read operations take place, separated from user data fields where R/W/E operations are allowed to be executed), the probe tips can then perform R/W/E operations on user data field in an array operation simultaneously. The simultaneous parallel operations of a large number of probes boost the data access speed tremendously.

The nanometer-wide tips of the probes perform the R/W/E operations by altering the surface physics of the polymer storage medium via (i) thermal [11], (ii) electric [13], or even (iii) magnetic [14] properties on a small dedicated region. The polymer storage medium is bistable and bonded on the micro X–Y stage or scanner platform during fabrication. The interference between adjacent bits must be kept to a minimal with high retention of the states after R/W/E operations to safeguard the reliability and integrity of the written-in user data. For batch fabrication, small form factor, and low cost, it is desirable for the micro X–Y stage to be fabricated using lithography processes. The micro X–Y stage with MEMS capacitive comb-driven microactuators should move the recording platform with a fast response while maintaining small mechanical crosstalk (axial coupling).

In the next sections, we give an introduction to a primary actuator and pay more attention to the various types of microactuators.

1.3 Actuators

1.3.1 Primary Actuator

Without loss of generality, the VCM actuator is taken as the primary actuator in this book. It is driven by a linear DC motor with restricted movement. In the application to hard disk drives, as seen in Figure 1.1, it contains a coil that is rigidly attached to the structure to be moved and suspended in a magnetic field created by permanent magnets. The VCM actuator moves in and out along a disk radius in one direction. When current passes through the coil, a force is produced which accelerates the actuator radially inward or outward, depending on the direction of the current. The produced force is a function of the current i_c. Approximately,

$$f_m = k_t i_c \tag{1.1}$$

where k_t is a linearized nominal value called torque constant. The resonance of the actuator is mainly due to the flexibility of the pivot bearing, arm, suspension, etc. When the bandwidth of a control loop is very low and the resonance may not be a limiting factor to the control design, the actuator model

Mechanical Actuation Systems 5

can be considered as the simplified and rigid one, which is a double integrator as follows:

$$P_v(s) = \frac{k_{vcm}}{s^2} \tag{1.2}$$

When the bandwidth is high, the actuator resonances have to be considered in the control design, since the flexible resonance modes will reduce the system stability and affect the control performance if ignored. Then, the actuator model becomes

$$P_v(s) = \frac{k_{vcm}}{s^2} P_r(s) \tag{1.3}$$

which includes the resonance model

$$P_r(s) = \prod_{i=1}^{N} P_{ri}(s) \tag{1.4}$$

and the resonance $P_{ri}(s)$ can be represented by any of the following forms:

$$P_{ri}(s) = \frac{\omega_i^2}{s^2 + 2\xi_i \omega_i s + \omega_i^2} \tag{1.5}$$

$$P_{ri}(s) = \frac{b_{1i}\omega_i s + b_{0i}\omega_i^2}{s^2 + 2\xi_i \omega_i s + \omega_i^2} \tag{1.6}$$

or

$$P_{ri}(s) = \frac{b_{2i}s^2 + b_{1i}\omega_i s + b_{0i}\omega_i^2}{s^2 + 2\xi_i \omega_i s + \omega_i^2} \tag{1.7}$$

where $\omega_i = 2\pi f_i$ ($i = 1, 2, \ldots, N$) corresponds to a single resonance frequency f_i and ξ_i is the associated damping coefficient. Equations (1.5) and (1.6) include zeros to facilitate a phase lift, which is usually associated with the resonance mode.

In addition to Equation (1.3), the VCM plant model can also be expressed in the summation format as follows:

$$P_v(s) = P_0(s) + \sum_{i=1}^{N} P_{ri}(s) \tag{1.8}$$

where $P_0(s)$ stands for the rigid part and the other term in the equation represents N flexible modes at the frequencies $\omega_i = 2\pi f_i$ ($i = 1, 2, 3, \ldots, N$).

The details about the VCM modeling and state-space description can be seen in [15].

1.3.2 Secondary Actuators

Multistage actuation systems use secondary actuators as fine positioners to increase the positioning accuracy. In hard disk drives for example, the secondary actuator piggyback on a VCM actuator is driven jointly with the VCM actuator through suspension [1,2], slider [16], or head [17–19]. There are mainly two types of secondary actuators which have received lots of interest in both academic areas and practical applications. One is the actuated suspension type [1,2], namely milliactuator. The other is the slider-driven type, i.e., microactuator, which is attached to the slider and moves the slider relatively to the suspension [16]. The latter is a collocated actuator with the read/write head or the sensor, and thus beneficial to attain better performance through control system design.

- PZT milliactuator

An example of PZT milliactuators is shown in Figure 1.3, where two PZT strips work in a pull–push format as one expands and the other shrinks. It is generally characterized by a Pade-delay and a two-pole roll-off model:

$$\text{padedelay} \times O(s^2) \tag{1.9}$$

Figure 1.4 shows the frequency responses of a PZT-actuated suspension, from which the transfer function in Equation (1.10) is obtained by curve-fitting to the measured frequency response. Notice that the PZT

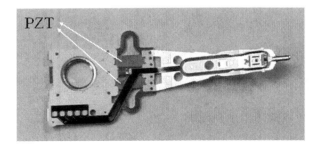

FIGURE 1.3
A PZT-actuated suspension. (From Ref. [15].)

Mechanical Actuation Systems

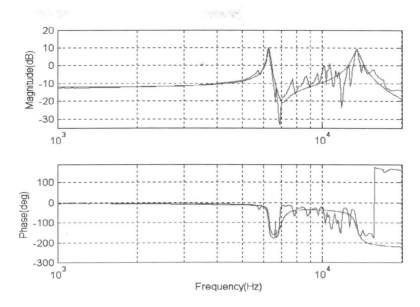

FIGURE 1.4
Frequency responses of a PZT-actuated suspension (smooth line: modeled; rough line: measured). (From Ref. [20].)

milliactuator is represented using a Pade-delay with two second-order resonance terms.

$$P_p(s) = \frac{-1438572836.9094(s - 6.157 \times 10^5)(s^2 + 923.7s + 1.934 \times 10^9)}{(s + 6.157 \times 10^5)(s^2 + 791.7s + 1.567 \times 10^9)(s^2 + 5089s + 7.195 \times 10^9)}$$

(1.10)

- Microactuators

There are three popular types of microactuators: electrostatic moving-slider microactuator [16], PZT slider-driven microactuator [19,21], and thermal microactuator [22–24]. Their basic transfer functions are included in Table 1.1, where their response times and main resonance frequencies are compared as well. It is known that the PZT type has the fastest response time and is the most easily controlled, while the thermal type responds very slowly. As for resonance frequency, the electrostatic type has the lowest resonance frequency, which is only 1–2 kHz, while the PZT type can have a very high resonance frequency and can be even as high as 30 kHz.

In hard disk drives, to realize high areal density and fast data transfer rate, it is needed to reduce track width continuously, and thus read/write head is required to stay at the track centers at a nanometer-level positioning accuracy, even though it faces a fast-rotating spindle motor. The dual-stage actuation

TABLE 1.1
Performance Comparison of the Microactuators

	Electrostatic [16]	PZT [19,21]	Thermal [22–25]
Basic transfer function	$\dfrac{K}{s^2+2\xi\omega s+\omega^2}$, ω: 1–2 kHz, $0<\xi<1$	$\dfrac{K}{s^2+2\xi\omega s+\omega^2}$, ω: >20 kHz, $0<\xi<1$	$\dfrac{K}{\tau s+1}$, τ: time constant
Response time (ms)	<0.1	<0.05	>0.1
Main resonance frequency (kHz)	1–2	20–25	>15

Source: Reprinted, with permission, from IEEE Trans. on Magnetics, 49(3), pp. 1082–1087, 2013.

system employing the secondary actuators has been applied in practice and proved to be an effective system that has satisfied the requirement. Their features of small size and high performance have produced more challenges in control systems. To all secondary actuators, an enough stroke is important to obtain a high servo control bandwidth, although it has trade-off relationship with resonance frequency or response time.

1.4 Single-Stage Actuation Systems

The single-stage actuation system employs the VCM actuator representatively. Its typical closed-loop control system is shown in Figure 1.5, where $P_v(s)$ and $C(z)$ represent the actuator system and its controller, respectively. There are practically effective solutions provided in literature for various problems such as fast and smooth tracking [26], error optimization, uncertainty [27],

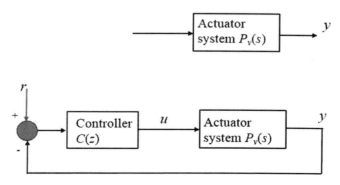

FIGURE 1.5
Block diagram of a single-stage actuation system and its typical closed-loop control system.

Mechanical Actuation Systems

and friction nonlinearity [28], and both classic and advanced control strategies together with algorithms have been extensively studied for the VCM actuator working as a single-stage system.

1.5 Dual-Stage Actuation Systems

In hard disk drives, the head positioning error with respect to the target track center needs to be made as small as possible in order to increase track density for a high storage capacity. There are various error sources mainly coming from (1) torque disturbance from spindle motor; (2) actuator pivot friction; (3) airflow-induced, non-repeatable turbulence to disk, suspension, and slider; (4) mechanical resonances; and (5) head sensing and electronic noise, media noise, and quantization noise. The influence of these error sources on the head positioning error must be minimized through the design of actuation mechanism and control system.

Dual-stage actuation mechanism consisting of a VCM actuator and a secondary actuator placed between the VCM and the sensor head has been extensively applied as an effective way to attain sufficiently high servo bandwidth and achieve the required disturbance and runout rejection. The VCM actuator is used as the primary stage to provide long track seeking but with poor accuracy and slow response time, while the secondary stage actuator is used to provide higher positioning accuracy and faster response but with a stroke limit. Figure 1.6 shows the block diagram of the dual-stage actuation system, where the overall position y comes from the combination of the two actuators' responses, y_v and y_p.

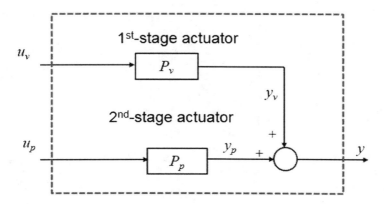

FIGURE 1.6
Block diagram of dual-stage actuation systems ($y = y_v + y_p$).

1.6 Three-Stage Actuation Systems

As understood from the briefing in Section 1.2, the microactuators are more collocated than the PZT-actuated suspension. However, due to the limited allowed deflection on the microactuator size and hence the allowable stroke relative to the disturbance, the control bandwidth is limited and the microactuator loop gain has to be restricted, which has limited the dual-stage control system's disturbance rejection capability and has a gain close to 0 dB at frequencies near the bandwidth (0 dB crossover frequency), regardless of having achieved a higher servo bandwidth [5], as shown in Figure 1.7. This suggests that adding more stroke to the microactuator will give more freedom to the servo loop shape, additionally to the traditional wisdom of just increasing its resonance frequency. To meet the requirements of less stroke constraint and saturation, a three-stage actuation system is thereby introduced for the demand of higher bandwidth. The block diagram of the three-stage actuation system is shown in Figure 1.8, where the overall position y is determined by the combination of the three actuators.

In a typical three-stage actuation system, the VCM actuator as the first-stage actuator works as the primary actuator, and PZT milliactuator as the second-stage actuator is used to strengthen the actuation system with a reasonable bandwidth. A third-stage actuator which is more collocated is then

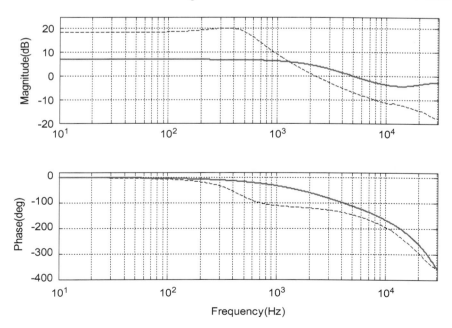

FIGURE 1.7
Open-loop frequency responses. (From Ref. [29].)

Mechanical Actuation Systems

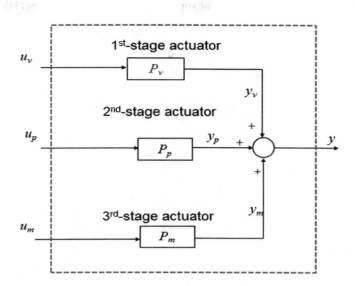

FIGURE 1.8
Block diagram of three-stage actuation systems ($y = y_v + y_p + y_m$).

needed to further push the bandwidth and regain the discounted bandwidth due to the VCM actuator. In Chapter 4 of this book, a thermal microactuator [25–30] is used as the third-stage actuator, and the details about system configuration and control system design are given there. It is worthwhile to mention that the dual-stage system control design is basically important to the control of the three-stage or multi-stage actuation systems.

References

1. K. Mori, T. Munemoto, H. Otsuki, Y. Yamaguchi, and K. Akagi, A dual-stage magnetic disk drive actuator using a piezoelectric device for a high track density, *IEEE Trans. Magn.*, 27(6), pp. 5298–5300, 1991.
2. R. B. Evans, J. S. Griesbach, and W. C. Messner, Piezoelectric microactuator for dual stage control, *IEEE Trans. Magn.*, 35(2), pp. 977–982, 1999.
3. A. Sinno, P. Ruaux, L. Chassagne, S. Topcu, Y. Alayli, G. Lerondel, S. Blaize, A. Bruyant, and P. Royer, Enlarged atomic force microscopy scanning scope: novel sample-holder device with millimeter range, *Rev. Sci. Instrum.*, 78(9), pp. 1–7, 2007.
4. M. Gauthier, and E. Piat, Control of a particular micro-macro positioning system applied to cell micromanipulation, *IEEE Trans. Autom. Sci. Eng.*, 3(3), pp. 264–271, 2006.
5. M. Rakotondrabe, and I. A. Ivan, Development and force/position control of a new hybrid thermo-piezoelectric microgripper dedicated to micromanipulation tasks, *IEEE Trans. Autom. Sci. Eng.*, 8(4), pp. 824–834, 2011.

6. G. Whitesides, and H. Love, The art of building small, *Sci. Am.*, 285(3), pp. 39–47, 2001.
7. Y. Michellod, P. Mullhaupt, and D. Gillet, Strategy for the control of a dual-stage nano-positioning system with a single metrology, *2006 IEEE Conference on Robotics, Automation and Mechatronics*, Bangkok, Thailand, pp. 1–8, 2006.
8. Y. Song, J. Wang, K. Yang, W. Yin, and Y. Zhu, A dual-stage control system for high-speed, ultra-precise linear motion, *Int. J. Adv. Manuf. Technol.*, 48(5), pp. 633–643, 2010.
9. L. Wang, J. Zheng, and M. Fu, Optimal preview control of a dual-stage actuator system for triangular reference tracking, *IEEE Trans. Control Syst. Technol.*, 22(6), pp. 2408–2416, 2014.
10. C. K. Pang, Y. Lu, C. Li, J. Chen, H. Zhu, J. Yang, J. Mou, G. Guo, B. M. Chen, and T. H. Lee, Design, fabrication, sensor fusion, and control of a micro X-Y stage media platform for probe-based storage systems, *Mechatronics*, 19(7), pp. 1158–1168, 2009.
11. E. Eletheriou, T. Antonakopoulos, G. K. Binning, G. Cherubini, M. Despont, A. Dholakia, U. Durig, M. A. Lantz, H. Pozidis, H. E. Rothuizen, and P. Vettiger, Millipede-A MEMS-based scanning-probe data-storage system, *IEEE Trans. Magn.*, 39(2), pp. 938–945, 2003.
12. M. S.-C. Lu, and G. K. Fedder, Position control of parallel-plate microactuators for probe-based data storage, *J. Microelectromech. Syst.*, 13(5), pp. 759–769, 2004.
13. H. Park, J. Jung, D.-K. Min, S. Kim, S. Hong, and H. Shin, Scanning resistive probe microscopy: imaging ferroelectric domains, *Appl. Phys. Lett.*, 84(10), pp. 1734–1736, 2004.
14. L. R. Carley, J. A. Main, G. K. Fedder, D. W. Greve, D. F. Guillou, M. S. C. Lu, T. Mukherjee, S. Santhanam, L. Abelmann, and S. Min, Single-chip computers with microelectromechanical systems-based magnetic memory, *J. Appl. Phys.*, 87(9), pp. 6680–6685, 2000.
15. C. Du, L. Xie, *Modeling and Control of Vibration in Mechanical Systems*, CRC Press, Boca Raton, 2010.
16. T. Hirano, M. White, H. Yang, K. Scott, S. Pattanaik, S. Arya, and F. Huang, A moving-slider MEMS actuator for high-bandwidth HDD tracking, *IEEE Trans. Magn.*, 40(4), pp. 3171–3173, 2004.
17. B. H. Kim and K. Chun, Fabrication of an electrostatic track-following microactuator for hard disk drive, *J. Micromech. Microeng.*, 11(1), pp. 1–6, 2001.
18. H. Fujita, K. Suzuki, M. Ataka, and S. Nakamura, A microactuator for head positioning system of hard disk drives, *IEEE Trans. Magn.*, 35(2), pp. 1006–1010, 1999.
19. K. Kurihara, M. Hida, S. Umemiya, M. Kondo, and S. Koganezawa, Rotating symmetrical piezoelectric microactuators for magnetic head drives, *Jpn. J. Appl. Phys.*, 45(9B), pp. 7471–7474, 2006.
20. C. Du, and G. Guo, Lowering the hump of sensitivity functions for discrete-time dual-stage systems, *IEEE Trans. Control Syst. Technol.*, 13(5), pp. 791–797, 2005.
21. T. Hirano, H. Takahashi, S. Hagiya, N. Nishiyama, and T. Tsuchiya, High bandwidth micro electro mechanical systems (MEMS) microactuator for hard-disk drive dual-stage tracking servo, *ASME Information Storage and Processing Systems Conf.*, Santa Clara, CA, June 14–15, 2010.

Mechanical Actuation Systems

22. S. Choe, H. Tomotsune, T. Eguchi, M. Kanamaru, T. Aono, S. Nakamura and A. Koide, A thermal driven microactuator for hard disk drive, *Digest of Asia-Pacific Magnetic Recording Conf.* 2010, Singapore, AB2, 2010.

23. J. A. Bain, A. El-Ghazaly, D. Bromberg, G. K. Fedder, and W. C. Messner, Electrothermal actuators for hard disk drive applications, *Digest of Asia-Pacific Magnetic Recording Conf.* 2010, Singapore, FB2, 2010.

24. K. Ono, H. Otsuki, J. Liu, J. Xu, and T. Arisaka, Thermal positioning control actuator for future high track density HDD, *ASME Information Storage and Processing Systems Conf.*, Santa Clara, CA, June 13–14, 2011.

25. G. K. Lau, J. Yang, B. Thubthimthong, N. B. Chong, C. P. Tan, and Z. He, Fast electrothermally activated micro-positioner using a high-aspect-ratio micro-machined polymeric composite, *Appl. Phys. Lett.*, 101, p. 033108, 2012.

26. Y. Li, V. Venkataramanan, G. Guo, and Y. Wang, Dynamic nonlinear control for fast seek-settling performance in hard disk drives, *IEEE Trans. Ind. Electron.*, 54(2), pp. 951–962, 2007.

27. C. Du, L. Xie, J. N. Teoh, and G. Guo, An improved mixed H2/H-infinity control design for hard disk drives, *IEEE Trans. Control Syst. Technol.*, 13(5), pp. 832–839, 2005.

28. C. Du, L. Xie, and J. Zhang, Compensation of VCM actuator pivot friction based on an operator modeling method, *IEEE Trans. Control Syst. Technol.*, 18(4), pp. 918–926, 2010.

29. C. Du, C. P. Tan, and J. Yang, Three-stage control for high servo bandwidth and small skew actuation, *IEEE Trans. Magn.*, 51(1), p. 310017, 2015.

30. J. Yang, G. K. Lau, C. P. Tan, N. B. Chong, B. Thubthimthong, L. Gonzaga, and Z. He, Silicon-polymer composite electro-thermal microactuator for high track density HDD, *ASME-ISPS/JSME-IIP Joint International Conf. on Micromechatronics for Information and Precision Equipment*, Santa Clara University, CA, US, 18–20, June 2012, pp. 66–68.

2

High-Precision Positioning Control of Dual-Stage Actuation Systems

2.1 Introduction

As stated in Chapter 1, the dual-stage actuation system is employed as an effective way to increase the overall control system bandwidth and achieve required disturbance rejections, so that high positioning accuracy can be accomplished. These are the objectives to be kept in mind when the controller design is being carried out. Examples of existing work are [1] on fundamental control design with concentration on pushing bandwidth [2], on phase-stabilized control to deal with narrow-band vibrations [3], on optimal control with the disturbance models, and [4] on a nonlinear method to improve system stability with the presence of secondary actuator's saturation.

The sensitivity function of the closed-loop system has provided a straight-forward view on its disturbance rejection capability. It is demanded that the sensitivity function magnitude in low-frequency range be sufficiently low, while its hump in high-frequency range needs to be low enough as well. However, this is difficult for controller design due to Bode integral constraint [5,6]. In view of this, the controller design for the dual-stage actuation systems adopts a weighting function to shape the sensitivity function. The weighting function is adjusted properly according to the desired performance so that the design results in a satisfactory sensitivity function with the designed controllers.

2.2 Control Schemes

A popular control scheme for the dual-stage actuation system is the decoupled structure as shown in Figure 2.1 [7], with the controllers $C_v(z)$ and $C_p(z)$, respectively, for the primary voice coil motor (VCM) actuator $P_v(s)$ and the secondary actuator $P_p(s)$, and $\hat{P}_p(z)$ as the approximation of P_p to estimate y_p.

15

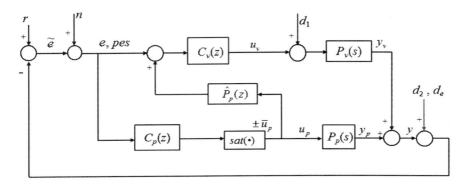

FIGURE 2.1
Decoupled control structure for the dual-stage actuation system with disturbances and noise injected.

In Figure 2.1, d_1 and d_2 denote internal disturbances, n is the measurement noise, and d_e stands for external vibration. For the application in hard disk drives, d_1 represents all torque disturbances and windage, d_2 represents disk and motor vibrations, n denotes the demodulation and measurement noise of the position error signal, and pes is the measured position error. The open-loop transfer function from pes to y is

$$G(z) = P_p(z)C_p(z) + P_v(z)C_v(z) + P_v(z)C_v(z)\hat{P}_p(z)C_p(z), \qquad (2.1)$$

and the overall sensitivity function of the closed-loop system from r to pes is

$$S(z) = \frac{1}{1+G(z)}, \qquad (2.2)$$

which is approximately

$$S(z) \approx \frac{1}{[1+P_p(z)C_p(z)][1+P_v(z)C_v(z)]} \qquad (2.3)$$

since within a certain high bandwidth

$$\hat{P}_p(z) \approx P_p(z).$$

The sensitivity functions of the VCM loop and the secondary actuator loop are, respectively, written as

$$S_v(z) = \frac{1}{1+P_v(z)C_v(z)}, \quad S_p(z) = \frac{1}{1+P_p(z)C_p(z)} \qquad (2.4)$$

High-Precision Positioning Control

then, Equation (2.3) means that

$$S(z) = S_v(z) \cdot S_p(z) \tag{2.5}$$

i.e., the dual-stage sensitivity function $S(z)$ is the product of $S_v(z)$ and $S_p(z)$. Thus, the dual-stage system control design can be decoupled into two independent controller designs: the VCM loop and the secondary actuator loop.

Let

$$G_v(z) = P_v(z)C_v(z) \tag{2.6}$$

and

$$G_p(z) = P_p(z)C_p(z) \tag{2.7}$$

then, the dual-stage open-loop transfer function (2.1) is approximately written as

$$G(z) = G_p(z) + G_v(z) + G_v(z)G_p(z). \tag{2.8}$$

With the disturbances d_1, d_2, d_e and noise n, the error $\tilde{e}$ and the measurement error e or pes with $r = 0$ are, respectively, contributed by the disturbances and noise through

$$\tilde{e} = -S(P_v d_1 + d_2 + d_e) - (1 - S)n \tag{2.9}$$

and

$$e = -S(P_v d_1 + d_2 + d_e) + Sn. \tag{2.10}$$

As usual, the disturbances and the noise are assumed to be independent of each other. Then, the power spectrum Φ_e of the error can be calculated by

$$\Phi_e = |P_v(f_k)S(f_k)|^2 |d_1(f_k)|^2 + |S(f_k)|^2 |d_2(f_k)|^2 + |S(f_k)|^2 |n(f_k)|^2 + |S(f_k)|^2 |d_e(f_k)|^2 \tag{2.11}$$

where f_k ($k = 1, 2, \ldots, N$) are frequency points. Equation (2.11) means that the disturbances and noise affect the error through the sensitivity function S and implies that the sensitivity function S plays an important role in the determination of the error magnitude.

Another type of control scheme is the parallel structure as shown in Figure 2.2. The open-loop transfer function from pes to y is

$$G(z) = P_p(z)C_p(z) + P_v(z)C_v(z) \tag{2.12}$$

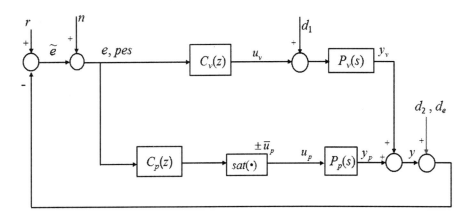

FIGURE 2.2
Parallel control structure for the dual-stage actuation system with disturbances and noise injected.

the overall sensitivity function of the closed-loop system from r to pes is

$$S(z) = \frac{1}{1+G(z)}$$

$$= \frac{1}{1+P_p(z)C_p(z)+P_v(z)C_v(z)}.$$

(2.13)

Obviously, the parallel structure and the decoupled structure can be converted to each other.

Because of the limited displacement range of the secondary actuator, the control efforts for the two actuators should be distributed properly when designing respective controllers to meet required performance, make the actuators not conflict with each other's control authority, as well as prevent the saturation of the secondary actuator. In the next sections, controller design methods will be presented in the continuous- and discrete-time domains.

2.3 Controller Design Method in the Continuous-Time Domain

H_∞ loop shaping method is used to design the controllers for the primary and secondary actuators to make the closed-loop system stable and meanwhile have a satisfactory high bandwidth, i.e., open-loop 0 dB crossover frequency.

The structure for the H_∞ loop shaping method is plotted in Figure 2.3, where $W(s)$ is a weighting function relevant to the designed control system performance such as the sensitivity function.

For a plant model $P(s)$, a controller $C(s)$ is to be designed such that the closed-loop system is stable and

$$\|T_{zw}\|_\infty < 1 \qquad (2.14)$$

is satisfied, where T_{zw} is the transfer function from w to z, i.e., $S(s)W(s)$. Equation (2.14) means that the sensitivity function $S(s)$ can be shaped similarly to the inverse of the chosen weighting function $W(s)$. One form of $W(s)$ is taken as

$$W(s) = \frac{\frac{1}{M}s^2 + 2\zeta\omega\frac{1}{\sqrt{M}}s + \omega^2}{s^2 + 2\omega\sqrt{\varepsilon}s + \omega^2\varepsilon} \qquad (2.15)$$

where ω is valued by the desired bandwidth, ε is used to determine the desired low frequency level of the resultant sensitivity function magnitude, and ζ is the damping ratio.

Associated with the weighting function $W(s)$, Figure 2.3 can be formulated as follows:

$$x(t) = Ax(t) + B_1 w(t) + B_2 u(t) \qquad (2.16a)$$

$$z(t) = C_1 x(t) + D_{11} w(t) + D_{12} u(t) \qquad (2.16b)$$

$$y(t) = C_2 x(t) + D_{21} w(t) + D_{22} u(t) \qquad (2.16c)$$

where

$$A = \begin{bmatrix} A_p & 0 \\ B_w C_p & A_w \end{bmatrix}, \; B_1 = \begin{bmatrix} 0 \\ B_w \end{bmatrix}, \; B_2 = \begin{bmatrix} B_p \\ B_w D_p \end{bmatrix}, \; C_1 = \begin{bmatrix} D_w C_p & C_w \end{bmatrix}$$

$$(2.17)$$

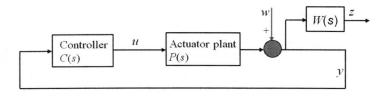

FIGURE 2.3
Block diagram for H_∞ loop shaping method to design the controller $C(s)$ with the weighting function $W(s)$.

$$D_{11} = D_w, \quad D_{12} = D_w D_p, \quad C_2 = \begin{bmatrix} C_p & 0 \end{bmatrix}, \quad D_{21} = 1, \quad D_{22} = D_p \quad (2.18)$$

(A_p, B_p, C_p, D_p) and (A_w, B_w, C_w, D_w) are, respectively, the state-space realization of plant $P(s)$ and weighting function $W(s)$.

Let (A_c, B_c, C_c, D_c) be the state-space description of $C(s)$. Then, (A_c, B_c, C_c, D_c) is to be designed such that Equation (2.14) is satisfied. The linear matrix inequality (LMI) approach stated in the following theorem is used to design the controller.

Theorem 2.1

All controllers of the form (A_c, B_c, C_c, D_c) such that $\|T_{zw}\|_\infty < \gamma$ holds are parameterized by the LMI [8]:

$$\begin{bmatrix} YA + A^T Y + VC_2 + C_2^T V^T & * & * & * \\ U^T + A + B_2 D_c C_2 & AX + X^T A^T + B_2 Z + Z^T B_2^T & * & * \\ B_1^T Y + D_{21}^T V^T & B_1^T + D_{21}^T D_c^T B_2^T & -\gamma I & * \\ D_{12} D_c C_2 + C_1 & C_1 X + D_{12} Z & D_{11} + D_{12} D_c D_{21} & -\gamma I \end{bmatrix} < 0$$

$$(2.19)$$

where * denotes an entry that can be deduced from the symmetry of the LMI, and the matrices Z, U, V, D_c and the symmetric matrices X and Y are the variables. A feasible controller is then given by choosing Ξ and Λ nonsingular such that $\Xi \Lambda = I - YX$ and calculating

$$D_c = D_c, C_c = (Z - D_c C_2 X)\Lambda^{-1} \quad (2.20)$$

$$B_c = \Xi^{-1}(V - YB_2 D_c) \quad (2.21)$$

$$A_c = \Xi^{-1}[U - Y(A + B_2 D_c C_2)X - YB_2 C_c \Lambda - \Xi B_c C_2 X]\Lambda^{-1} \quad (2.22)$$

The LMI (2.19) can be solved using the Matlab LMI toolbox, and subsequently the controller can be obtained from (2.20)–(2.22) with the LMI solutions. The obtained controller is in the continuous form and needs to be discretized before being practically implemented in the control system.

There always exists a minimum level γ that makes the LMI (2.19) solvable, which gives a sensitivity function more similar to the inverse of the weighting function than a larger γ. The solvability of the LMI is also related to the chosen weighting $W(s)$, which, although various, must be realistic due to the Bode limitation [5,6].

The primary and secondary actuator control loops are designed separately for the dual-stage control systems. But when designing their respective controllers, certain performances are required for the two actuators, so that the control efforts for the two actuators are distributed properly and the actuators don't conflict with each other's control authority. As seen in Figure 2.4, the VCM primary actuator open loop has a higher gain at low frequencies, and the secondary actuator open loop has a higher gain in the high-frequency range. The sensitivity functions are shown in Figure 2.5, where the hump of S_v is arranged within the frequency band of S_p below 0 dB, and the high-frequency hump of S_p is lowered as much as possible. This needs to decrease the bandwidth of the primary actuator loop and increase the bandwidth of the secondary actuator loop, which can be realized by adjusting the weighting functions $W_v(s)$ for the primary actuator and $W_p(s)$ for the secondary actuator. A low-hump dual-stage sensitivity function can also be achieved by appropriately adjusting the weighting functions [10].

Alternatively, the dual-stage system control can be formulated as a MISO (multi-input-single-output) problem. However, to meet the specific requirements for the two actuators, the controller design through solving the MISO problem would be more complicated.

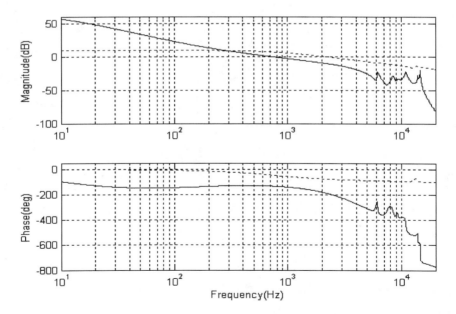

FIGURE 2.4
Frequency responses of $G_v(s) = C_v(s)P_v(s)$ (solid line) and $G_p(s) = C_p(s)P_p(s)$ (dotted line). (From Ref. [12].)

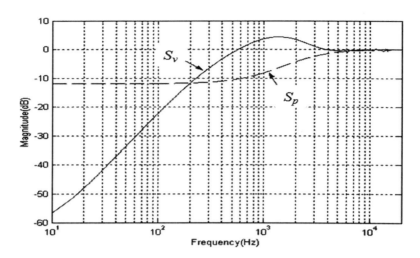

FIGURE 2.5
Frequency responses of $S_v(s)$ and $S_p(s)$. (From Ref. [12].)

2.4 Controller Design Method in the Discrete-Time Domain

The discrete-time H_∞ loop shaping method is another choice for the controller designs of the primary and secondary actuator loops. The structure of the discrete-time H_∞ loop shaping method is the same as in Figure 2.3 with $C(s)$ and $W(s)$ replaced with $C(z)$ and $W(z)$, respectively.

In the discrete-time domain, Figure 2.3 is formulated as follows:

$$x(k+1) = Ax(k) + B_1 w(k) + B_2 u(k) \tag{2.23}$$

$$z(k) = C_1 x(k) + D_{11} w(k) + D_{12} u(k) \tag{2.24}$$

$$y(k) = C_2 x(k) + D_{21} w(k) + D_{22} u(k) \tag{2.25}$$

where

$$A = \begin{bmatrix} A_p & 0 \\ B_w C_v & A_w \end{bmatrix}, \quad B_1 = \begin{bmatrix} 0 \\ B_w \end{bmatrix}, \quad B_2 = \begin{bmatrix} B_p \\ B_w D_p \end{bmatrix} \tag{2.26}$$

$$C_1 = \begin{bmatrix} D_w C_p & C_w \end{bmatrix}, \quad D_{11} = D_w, \quad D_{12} = D_w D_p, \quad C_2 = \begin{bmatrix} C_p & 0 \end{bmatrix}, \quad D_{21} = 1, \quad D_{22} = D_p \tag{2.27}$$

High-Precision Positioning Control

$\begin{bmatrix} A_p & B_p & C_p & D_p \end{bmatrix}$ and $\begin{bmatrix} A_w & B_w & C_w & D_w \end{bmatrix}$ are, respectively, the state-space realizations of the plant $P(z)$ and the weighting function $W(z)$. The LMI approach stated in the following theorem is used to design a controller such that

$$\left\| T_{zw} \right\|_\infty < 1 \tag{2.28}$$

is satisfied, where T_{zw} is the transfer function from w to z, i.e., $S(z)W(z)$.

Theorem 2.2

All controllers of the form (A_c, B_c, C_c, D_c) such that $||T_{zw}||_\infty < \gamma$ holds are parameterized by the following LMI [9,10]:

$$\begin{bmatrix} -Y & * & * & * & * & * \\ -I & -X & * & * & * & * \\ YA + VC_2 & U & -Y & * & * & * \\ A + B_2 D_c C_2 & AX + B_2 Z & -I & -X & * & * \\ C_1 + D_{12} D_c C_2 & C_1 X + D_{12} Z & 0 & 0 & -I & * \\ 0 & 0 & B_1^T Y + D_{21}^T V^T & B_1^T + D_{21}^T D_c^T B_2^T & D_{11}^T + D_{21}^T D_c^T D_{12}^T & -\gamma^2 I \end{bmatrix} < 0 \tag{2.29}$$

where $*$ denotes an entry that can be deduced from the symmetry of the LMI, and the matrices Z, U, V, D_c and the symmetric matrices X and Y are the variables.

A feasible controller is then given by choosing Ξ and Λ nonsingular such that $\Xi\Lambda = I - YX$ and calculating

$$C_c = (Z - D_c C_2 X)\Lambda^{-1} \tag{2.30}$$

$$B_c = \Xi^{-1}(V - YB_2 D_c) \tag{2.31}$$

$$A_c = \Xi^{-1}[U - Y(A + B_2 D_c C_2)X - YB_2 C_c\Lambda - \Xi B_c C_2 X]\Lambda^{-1} \tag{2.32}$$

The designed controller resulting from (2.30)–(2.32) is already in discrete-time form, and thus discretization is not necessary [10].

Remark 2.1

In order to have more freedom in the loop shaping for the secondary actuator loop, a higher-order weighting function can be used and given by

$$W(s) = W_m(s) \cdot W_c(s) \tag{2.33}$$

with $W_c(s)$ as in (2.15), and

$$W_m(s) = \frac{s^2 + 2\xi_{m1}\omega_m\sqrt{\varepsilon_m}\, s + \omega_m^2 \varepsilon_m}{s^2 + 2\xi_{m2}\omega_m s + \omega_m^2} \tag{2.34}$$

where $\omega_m < \omega$, $\xi_{m1} < 1$, and $\xi_{m2} < 1$. To illustrate the weighting functions in the frequency domain, their inverse frequency responses are plotted in Figure 2.6, where, for $|W^{-1}(s)|$, the major difference from $|W_c^{-1}(s)|$ is in the low-frequency part due to $|W_m^{-1}(s)|$.

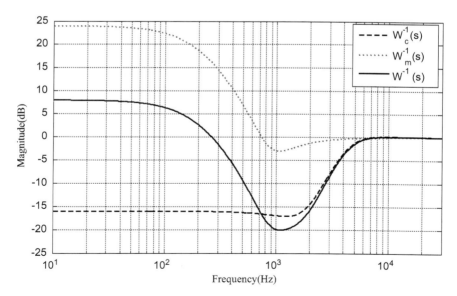

FIGURE 2.6
Frequency responses of $|W_c^{-1}(s)|$, $|W_m^{-1}(s)|$, and $|W^{-1}(s)|$.

High-Precision Positioning Control

Remark 2.2

It is known that there are typical traditional methods to design a controller for VCM actuators. Suppose that the dominant VCM resonant modes are compensated for by notch filters and the compensated model can be approximated by a double integrator in the frequency range of interest. A typical PID (proportional-integral-derivative) controller or a lag-lead compensator recommended in [4] is then used as the VCM controller. This design methodology uses the lag portion of the controller to increase the low frequency gain for low frequency disturbance rejection and positioning accuracy. The lead portion of the controller increases phase margin to ensure stability in the crossover region. This design also requires the designer to tune only one parameter α for the desired gain crossover frequency f_V. A slight modification is made, and the controller can take the following form [11]:

$$C_V(s) = K_V \frac{s + \dfrac{2\pi f_V}{2\alpha}}{s + 2\pi 10} \cdot \frac{s + \dfrac{2\pi f_V}{\alpha}}{s + 2\alpha 2\pi f_V} \tag{2.35}$$

with $5 < \alpha < 10$ used typically. K_V can be calculated by setting

$$\left| C_V(j2\pi f_V) G_V(j2\pi f_V) \right| = 1. \tag{2.36}$$

2.5 Application in the Controller Design of a Dual-Stage Actuation System

2.5.1 Actuator Modeling Based on Frequency Responses Measurement

The actuators are modeled by measuring their frequency responses with an LDV (laser Doppler vibrometer) for displacement measurement and a Dynamic Signal Analyzer for frequency response generation.

The frequency responses of the VCM actuator of the dual-stage actuation system are shown in Figure 2.7. And those of the PZT (Pb-Zr-Ti) secondary actuator are shown in Figure 1.3 with the transfer function in (1.10). The transfer functions are obtained by doing curve-fitting to the measured frequency responses.

2.5.2 Controller Design and Simulation

The used sampling rate for the control system is 40 kHz. The discrete-time models of the actuators are obtained by discretizing using the "zero-order-hold" method. The method in Section 2.4 is used to design the controllers

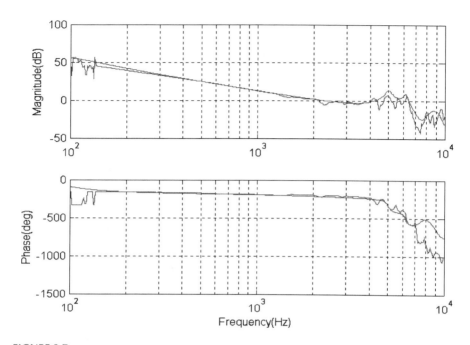

FIGURE 2.7
VCM actuator frequency responses (smooth line: modeled; rough line: measured). (From Ref. [10].)

$C_v(z)$ and $C_p(z)$. The adopted control structure is the parallel structure in Figure 2.2.

A basic requirement of the dual-stage actuation control system is to make the individual primary and secondary actuator loops stable. It also requires that the primary actuator path has a higher gain than the secondary actuator path at low-frequency range and the secondary actuator path has a higher gain than the primary actuator path in high-frequency range. These can be achieved by choosing appropriate weighting functions for the controllers' design.

With $\omega = 2\pi 350$, $\varepsilon = 10^{-3.2}$, $\zeta = 0.4$ in (2.15), a controller $C_v(z)$ for the primary actuator is designed, as shown in Figure 2.8 using the LMI approach in Section 2.4. Applying the LMI approach, the controller $C_p(z)$ for the milliactuator (1.10) is designed and shown in Figure 2.9 with $\omega = 2\pi 3{,}700$, $\varepsilon = 10^{-1.38}$, $\zeta = 1$, and $M = 0.07^{1/2}$ in (2.15).

The frequency responses of the open-loop transfer functions $P_v(z)C_v(z)$ and $P_p(z)C_p(z)$ are shown in Figure 2.10 where it is clearly noticed that the above requirements have been met. The frequency responses of $G(z) = P_v(z)C_v(z) + P_p(z)C_p(z)$ are shown in Figure 2.14, from which it is known that the gain margin is 8.4 dB, the phase margin is 64°, and the bandwidth is 2.49 kHz. The magnitude of the sensitivity function $S(z) = [1 + G(z)]^{-1}$ of the overall dual-stage control system is shown in Figure 2.11, with the hump below 3 dB.

High-Precision Positioning Control

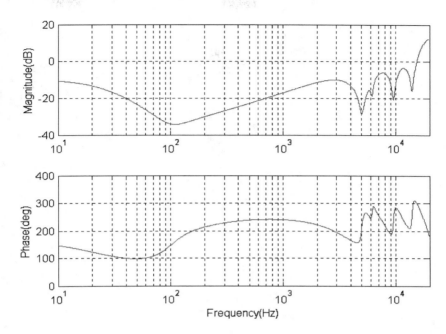

FIGURE 2.8
Frequency responses of the VCM controller $C_v(z)$. (From Ref. [10].)

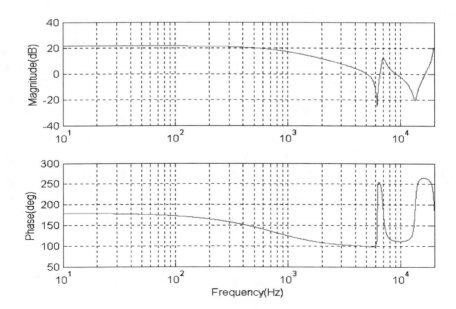

FIGURE 2.9
Frequency responses of the PZT controller $C_p(z)$. (From Ref. [10].)

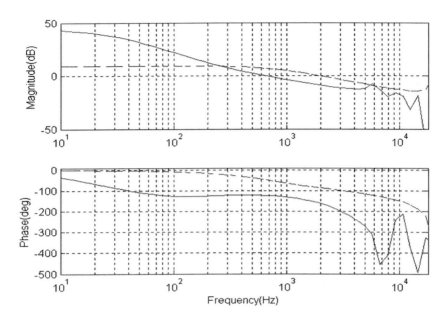

FIGURE 2.10
Frequency responses of $G_v(z) = C_v(z)P_z(z)$ (solid) and $G_p(z) = C_p(z)P_p(z)$ (dashed). (From Ref. [10].)

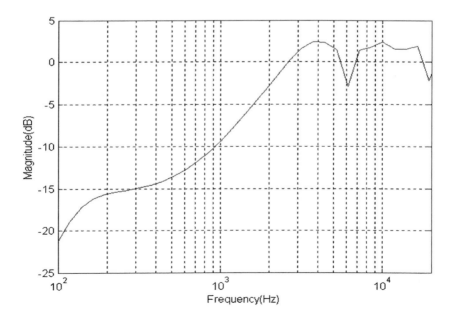

FIGURE 2.11
Sensitivity function of the dual-stage actuation control system. (From Ref. [10].)

High-Precision Positioning Control

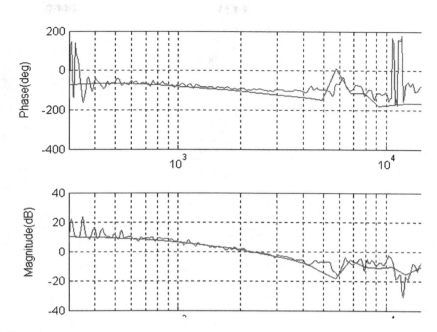

FIGURE 2.12
Frequency responses of the dual-stage open-loop transfer function $G(z) = P_v(z)C_v(z) + P_p(z)C_p(z)$ (smooth line: modeled; rough line: measured). (From Ref. [10].)

2.5.3 Experimental Results

A dSPACE together with an LDV [10] is used to implement the controllers for the dual-stage actuation system. The tested and simulated frequency responses of the dual-stage open-loop transfer function are shown in Figure 2.12. And, the sensitivity functions are shown in Figure 2.13, where the hump of the sensitivity function is lower than 3 dB, which is better than that with the PID control, as compared in Figure 2.13. As the hump is above 0 dB, the disturbances in the frequency range where the hump is will be amplified. Thus, a lower hump leading to a less amplification is desired.

The step response is seen in Figure 2.14, which shows that the closed-loop system is stable and follows the reference well in real time. Meanwhile, it shows the control signals of the VCM actuator in Channel 2 and the PZT milliactuator in Channel 3 as well.

2.6 Conclusion

In this chapter, the controller design has been discussed for high-precision positioning control of the dual-stage actuation systems. The H_∞ loop shaping

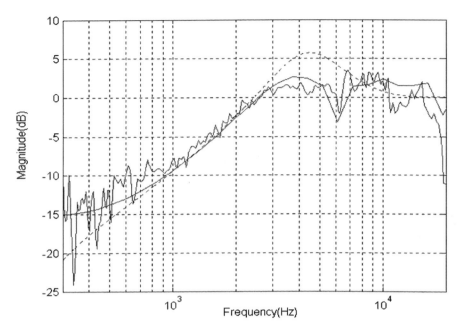

FIGURE 2.13
Magnitude of the sensitivity function $S(z)$ of the dual-stage actuation control system (smooth line: modeled; rough line: measured; dotted line: PID design). (From Ref. [10].)

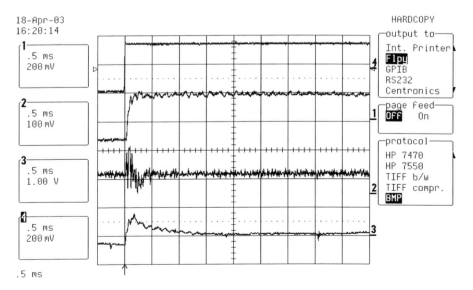

FIGURE 2.14
Step response of the dual-stage actuation control system. (From Ref. [10].)

High-Precision Positioning Control

method has been applied and the design method has been presented in both continuous and discrete-time domains. With the weighting functions, the desired sensitivity function can be achieved with the H_∞ loop shaping method. Such a design method can produce robust controllers with more disturbance rejection in the low frequency range and less disturbance amplification in the high-frequency range, and thus it is an effective method to achieve high positioning accuracy through loop shaping. The design method has been further used in the controller design of a dual-stage actuation system composed of a VCM primary actuator and a PZT milliactuator. And, it will be applied to the controller design of other types of dual-stage actuation systems and three-stage actuation systems in the later chapters.

References

1. S. K. Aggarwal, D. A. Horsley, R. Horowitz, and A. P. Pisano, Micro-actuators for high density disk drives, *Proceeding of the American Control Conference*, pp. 3979–3984, New Mexico, USA, 4–6 June, 1997.
2. M. Kobayashi, S. Nakagawa, and S. Nakamura, A phase-stabilized servo controller for dual-stage actuators in hard disk drives, *IEEE Trans. Magn.*, 39(2), pp. 844–850, 2003.
3. Z. Li, G. Guo, B. M. Chen, and T. H. Lee, An optimal track following design towards the highest track-per-inch in hard disk drives, *J. Inf. Stor. Process. Syst.*, 3, pp. 27–41, 2001.
4. G. Guo, D. Wu, and T. C. Chong, Modified dual-stage controller for dealing with secondary stage actuator saturation, *IEEE Trans. Magn.*, 39(6), pp. 3587–3592, 2003.
5. H. W. Bode, *Network Analysis and Feedback Amplifier Design*, Van Nostrand, Princeton, NJ, 1945.
6. B. Wu and E. Jonckheere, A simplified approach to Bode's theorem for continuous-time and discrete-time systems, *IEEE Trans. Autom. Control*, 37(11), pp. 1797–1802, 1992.
7. K. Mori, T. Munemoto, H. Otsuki, Y. Yamaguchi, and K. Akagi, A dual-stage magnetic disk drive actuator using a piezoelectric device for a high track density, *IEEE Trans. Magn.*, 27(6), pp. 5298–5300, 1991.
8. C. Scherer, P. Gahinet, and M. Chilali, Multiobjective output-feedback control via LMI optimization, *IEEE Trans. Autom. Control*, 42(7), pp. 896–911, 1997.
9. M. C. de Oliveira, J. C. Geromel, and J. Bernussou, An LMI optimization approach to multiobjective and robust H_∞ controller design for discrete-time systems, *Proceedings of the 38th IEEE Conf. on Decision and Control*, vol. 4, pp. 3611–3616, 1999.
10. C. Du and G. Guo, Lowering the hump of sensitivity functions of discrete-time dual-stage systems, *IEEE Trans. Control Syst. Technol.*, 13(5), pp. 791–797, 2005.

11. C. K. Pang, D. Wu, G. Guo, T. C. Chong, and Y. Wang, Suppressing sensitivity hump in HDD dual-stage servo systems, *Microsyst. Technol.*, 11, pp. 653–662, 2005.
12. C. Du, G. Guo, and D. Wu, Low-hump sensitivity function design for dual-stage HDD systems with different micro actuators, *IEE Proc. - Control Theory Appl.*, 152(6), pp. 655–661, 2005.

3

Control of Thermal Microactuator-Based, Dual-Stage Actuation Systems

3.1 Introduction

The thermal microactuator introduced in Chapter 1 has advantages such as large driving force, high natural frequency, and simple assembly, and thus is considered as a promising technology for accurate positioning. This chapter focuses on the control of the dual-stage actuation system using a thermal microactuator [1–3] as the secondary actuator. The microactuator-slider-suspension assembly is attached to a VCM (voice coil motor) arm, which constitutes the dual-stage actuation system. The detailed modeling of the thermal microactuator is also presented before the controller design. The decoupled control scheme discussed in Chapter 2 is adopted for the dual-stage system with the thermal microactuator. The controllers of the VCM and the thermal microactuator are designed with the H_∞ loop shaping method, which guarantee the stability and an extremely high bandwidth of the overall dual-stage system control loop.

3.2 Modeling of a Thermal Microactuator

A limitation of thermal microactuators is its slow response time, which hinders pushing control bandwidth. To obtain a high control bandwidth, an enough stroke is also important, although there is a trade-off relationship between response time and stroke length. Therefore, an optimal balance between the response time and the stroke is crucial for the thermal micro-actuator design.

A thermal microactuator is proposed in [1–3] with the mechanical design and actuation principle addressed in detail. It is used in the dual-stage actuation system as the secondary actuator, which is sandwiched between the slider and the suspension for application in hard disk drives and actually

33

is a slider-driven micro thermal actuator (MTA). Without loss of generality, the VCM actuator works as the primary actuator in this dual-stage actuation system.

The thermal microactuator model, denoted by $P_m(s)$, has two portions, namely $P_t(s)$ and $\tilde{P}(s)$, i.e.,

$$P_m(s) = P_t(s)\tilde{P}(s) \tag{3.1}$$

where $P_t(s)$ is modeled by the gain K and a first-order system with a time constant τ and is given by

$$P_t(s) = \frac{K}{\tau s + 1} \tag{3.2}$$

The mechanical part $\tilde{P}(s)$ consists of the resonance of the microactuator-slider assembly and the excited resonance modes of the suspension, and is expressed as

$$\tilde{P}(s) = \prod_{i=1}^{N} \frac{\omega_{i2}^2(s^2 + 2\zeta_{i1}\omega_{i1}s + \omega_{i1}^2)}{\omega_{i1}^2(s^2 + 2\zeta_{i2}\omega_{i2}s + \omega_{i2}^2)} \tag{3.3}$$

where ω_{i1} and ω_{i2} are the anti-resonance and resonance frequencies, respectively; ζ_{i1} and ζ_{i2} are the corresponding damping ratios; and N is the number of resonance modes.

The actuators' frequency responses are individually measured with the LDV (laser Doppler vibrometer) for displacement measurement and the equipment Dynamic Signal Analyzer with swept sine signal as excitation. The LDV range is set as $100\,\mathrm{nm/V}$.

The thermal microactuator is inherently driven by power. Figure 3.1 shows the measured frequency responses with different excitation levels, where the frequency responses don't change much for different excitation levels, which implies that the power driven thermal microactuator has a good linearity characteristic. This is advantageous for controller design and implementation. It is noted that if it is driven by voltage, the nonlinearity appears severely, which is observed in Figure 3.2.

For this thermal microactuator plant, the time constant $\tau = 0.5\,\mathrm{ms}$, the slider resonance is at $34\,\mathrm{kHz}$, and the stroke is $45\,\mathrm{nm}$ with an input of $3.5\,\mathrm{V}$. The resonance at $2.3\,\mathrm{kHz}$ in Figures 3.1 and 3.2 is the slider gimbal mode, which is supposed to disappear when the slider flies on disk. Moreover, the displacement versus input voltage at $100\,\mathrm{Hz}$ with $3.5\,\mathrm{V}$ excitation level is plotted in Figure 3.3, where the microactuator driven by power exhibits a good linearity behavior.

In addition, Figure 3.4 plots the frequency responses of the VCM actuator used in this dual-stage system. And, $P_v(s)$ is used to denote the VCM actuator plant.

Dual-Stage Actuation Systems

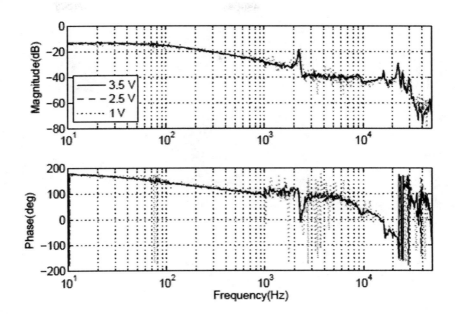

FIGURE 3.1
Frequency responses of MTA driven by power with different excitation levels. (From Ref. [4].)

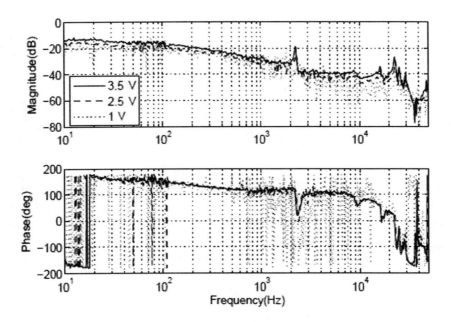

FIGURE 3.2
Frequency responses of the thermal microactuator driven by voltage with different excitation levels. (From Ref. [4].)

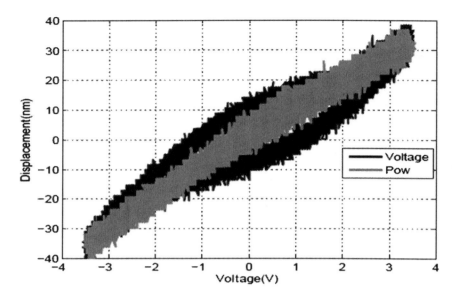

FIGURE 3.3
Displacement versus input of the voltage and the power driven microactuator. (From Ref. [4].)

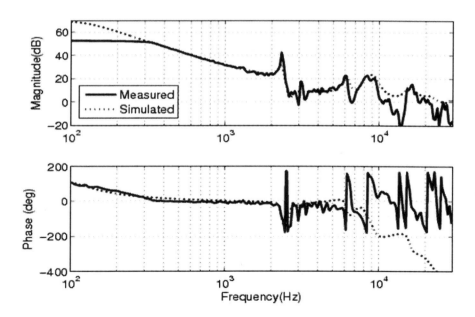

FIGURE 3.4
Frequency responses of the VCM plant. (From Ref. [4].)

3.3 Controller Design and Performance Evaluation

The decoupled master–slave structure is adopted for the control of the dual-stage actuation system and depicted in Figure 3.5. As shown in Chapter 2, the sensitivity function of the dual-stage control system is given by

$$S(z) \approx S_v(z)S_m(z) \quad (3.4)$$

where

$$S_v(z) = \frac{1}{1+P_v(z)C_v(z)} \quad (3.5)$$

$$S_m(z) = \frac{1}{1+P_m(z)C_n(z)C_m(z)} \quad (3.6)$$

which implies that the controller $C_v(z)$ for the VCM loop and $C_m(z)$ for the microactuator loop can be designed independently to shape the overall sensitivity function $S(z)$.

The VCM controller $C_v(z)$ is conventionally designed either with a PID (proportional-integral-derivative) method or with an advanced method such as the H_∞ loop shaping method stated in Chapter 2.

In Figure 3.5, $\hat{P}_t(s)$ is an approximated model of $P_m(s)$ cascaded with the pre-compensator $C_n(z)$. The pre-compensator is used to compensate for certain resonances, so that $P_m(z)C_n(z)$ can be approximated by a simple form of $\hat{P}_t(z)$. The microactuator model before and after the pre-compensation is compared in Figure 3.6, where it is seen that the approximated model $\hat{P}_t(z)$ is about the same as $P_m(z)C_n(z)$ within 4 kHz. Here, $\hat{P}_t(z)$ is designed as the

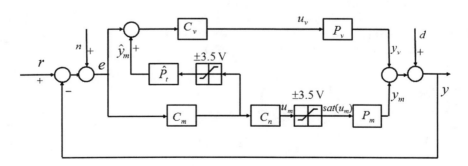

FIGURE 3.5
Control structure of the dual-stage system with the thermal microactuator. (From Ref. [4].)

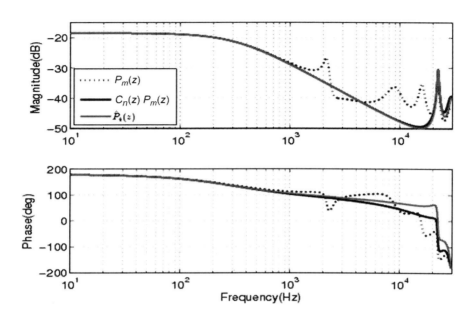

FIGURE 3.6
Frequency responses of the thermal microactuator before and after the pre-compensation. (From Ref. [4].)

discretized form of $K/(\tau s+1)$ cascaded with two resonance modes at 22 and 29 kHz.

The microactuator controller $C_m(z)$ is then designed for $P_m(z)C_n(z)$ or $\hat{P}_t(z)$ using the H_∞ loop shaping method with the framework shown in Figure 3.7, where $W_s(z)$ is a performance weighting function and $W_c(z)$ is a weighting function to limit control signal. A challenge for the microactuator controller design comes from the input constraint, which is more critical than the PZT milliactuator as its movement range is in nanometer level and is much smaller than the micrometer level of the milliactuator. Thus, the weighting $W_c(z)$ is employed in order to limit the control signal when meeting the required performance. Here, the input constraint is ±3.5 V.

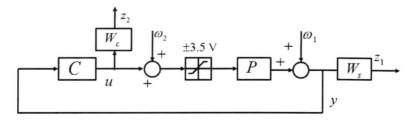

FIGURE 3.7
Block diagram of the H_∞ loop shaping for the MTA controller design. (From Ref. [4].)

Dual-Stage Actuation Systems

For the given plant $P(z) = \hat{P}_t(z) : (A_p, B_p, C_p, D_p)$ with the state x_p and the weighting functions $W_s(z) : (A_w, B_w, C_w, D_w)$ with the state x_w, the H_∞ loop shaping method associated with Figure 3.7 is formulated as follows:

$$x(k+1) = Ax(k) + B_1\omega(k) + B_2u(k) \tag{3.7}$$

$$z(k) = C_1x(k) + D_{11}\omega(k) + D_{12}u(k) \tag{3.8}$$

$$y(k) = C_2x(k) + D_{21}\omega(k) + D_{22}u(k) \tag{3.9}$$

where

$$A = \begin{bmatrix} A_p & 0 \\ B_wC_p & A_w \end{bmatrix}, \quad B_1 = \begin{bmatrix} 0 & B_p \\ B_w & B_wD_p \end{bmatrix}, \quad B_2 = \begin{bmatrix} B_p \\ B_wD_p \end{bmatrix}, \quad C_1 = \begin{bmatrix} D_wC_p & C_w \\ 0 & 0 \end{bmatrix} \tag{3.10}$$

$$D_{11} = \begin{bmatrix} D_w & D_wD_p \\ 0 & 0 \end{bmatrix}, \quad D_{12} = \begin{bmatrix} D_wD_p \\ W_c \end{bmatrix}, \quad C_2 = \begin{bmatrix} C_p & 0 \end{bmatrix}, \quad D_{21} = \begin{bmatrix} 1 & D_p \end{bmatrix}, \quad D_{22} = D_p \tag{3.11}$$

With this augmented system (3.7)–(3.9) with the state $x = \begin{bmatrix} x_p^T & x_w^T \end{bmatrix}^T$, a controller $C(z)$ is to be designed such that the closed-loop system is stable and $\|T_{z\omega}\|_\infty < \gamma$ for a given γ, where $T_{z\omega}$ is the transfer function from $\omega = \begin{bmatrix} \omega_1^T & \omega_2^T \end{bmatrix}^T$ to $z = \begin{bmatrix} z_1^T & z_2^T \end{bmatrix}^T$. The LMI (linear matrix inequality) approach stated in Theorem 2.2 is used to solve the controller $C(z)$ or $C_m(z)$ in this application.

Remark 3.1

The performance of $\|T_{z_1\omega_1}\| < 1$ means that the sensitivity function can be shaped similarly to the inverse of the chosen weighting function $W_s(z)$. $W_s(s)$ is chosen as in (2.15), and $W_s(z)$ is obtained by discretizing $W_s(s)$. $\|T_{z_2\omega_2}\| < \gamma$ means that the input signal is bounded by using the weighting function $W_c(z)$.

Note that the control scheme [4] in Figure 3.5 uses a saturation block $sat(u_t) = \pm 3.5\,\text{V}$ in the decoupling loop. This helps to improve the stability of the control loop against the microactuator saturation [5]. Without $sat(u_t)$ in the decoupling loop, when the microactuator is saturated for a sufficiently long time, the VCM loop tends to be potentially unstable, leading to the instability of the whole dual-stage system.

The sampling rate for the dual-stage system control is 60 kHz. The 15th-order controller C_v for the VCM actuator is designed using the H_∞ loop shaping method. The sixth-order notch filter $C_n(z)$ and the seventh-order controller $C_m(z)$ are designed in sequence for the MTA. The frequency responses of the cascaded controller $C_n(z)C_m(z)$ for the microactuator are shown in Figure 3.8. Note that $C_n(z)$ is designed to pre-compensate for the resonances at 2.3, 9.0, and 16 kHz, except for the higher-frequency resonances at 22 and 29 kHz. The notch at 2.3 kHz in Figure 3.8 is to deal with the mode of 2.3 kHz as observed in Figure 3.1. The relatively low gain in the low frequency range is to lower the control signal in order to avoid saturation and achieve high bandwidth performance as well.

The open-loop transfer functions of the VCM actuator, the MTA, and the overall dual-stage system are respectively given by

$$G_v(z) = P_v(z)C_v(z)$$
$$G_m(z) = P_m(z)C_n(z)C_m(z) \qquad (3.12)$$
$$G(z) = G_v(z) + G_m(z) + G_v(z)G_m(z).$$

The frequency responses of $G_v(z)$, $G_m(z)$, and $G(z)$ are shown in Figure 3.9. It is noticed that 580 Hz bandwidth, 12.6 dB gain margin, and 40.3° phase

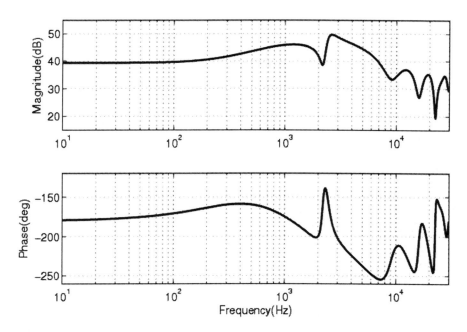

FIGURE 3.8
Frequency response of the thermal microactuator controller $C_n(z)C_m(z)$. (From Ref. [4].)

Dual-Stage Actuation Systems

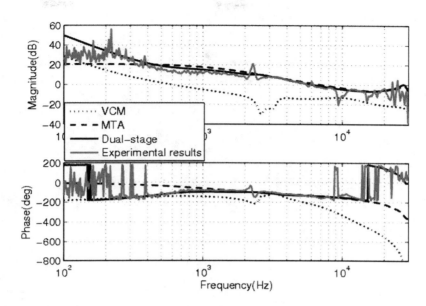

FIGURE 3.9
Open-loop frequency responses of the VCM loop, the MTA loop, and the dual-stage system. (From Ref. [4].)

margins are obtained with the VCM control. It should be mentioned that the bandwidth enhancement is limited by the resonance mode at 2.3 kHz as seen in the frequency responses in Figures 3.4 and 3.6. With the microactuator, 6 kHz bandwidth, 6.9 dB, and 44° stability margins are achieved by employing the dual-stage systems.

3.4 Experimental Results

To carry out the control testing, the LDV is used to measure the slider off-track displacement y with the range of 100 nm/V, and the controllers are implemented with dSPACE 1103. Figure 3.10 shows the experimental system structure.

The experimental results of the dual-stage open-loop frequency responses are shown in Figure 3.9. The sensitivity functions (3.5), (3.6), and (3.4) corresponding to the VCM actuator, the microactuator, and the overall dual-stage system are shown in Figure 3.11. The dual-stage sensitivity function means that the control system is capable of rejecting the disturbances with frequencies up to 5 kHz. For verification, we inject a disturbance d to the dual-stage system as an output disturbance. The power spectrum of the disturbance

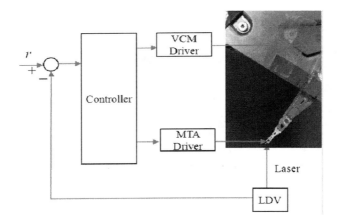

FIGURE 3.10
Experimental dual-stage actuation system with the thermal microactuator. (From Ref. [4].)

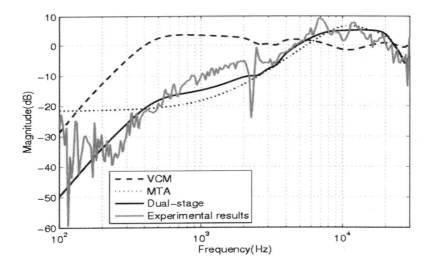

FIGURE 3.11
Sensitivity functions of the VCM loop, the microactuator loop, and the overall dual-stage system. (From Ref. [4].)

d and the error e is shown in Figure 3.12 with the peaks at 0.6, 1, 3, and 4 kHz. As observed from the spectrum of the error signal e in Figure 3.12, these peaks are apparently attenuated. The peaks after 5 kHz are due to the system measurement noise and the sensitivity function. All these can be easily understood by considering that the spectrum of the error e is given by $\Psi_e = |S(z)|^2 |d|^2 + |S(z)|^2 |n|^2$, where d is the output disturbance and n is the system measurement noise.

Dual-Stage Actuation Systems

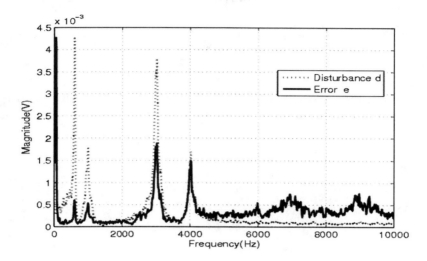

FIGURE 3.12
Power spectrum of disturbance d and error e (rejected the disturbance in the frequencies up to 4.5 kHz). (From Ref. [4].)

The seeking performance of the dual-stage system is investigated with a 40 nm step reference. As shown in Figure 3.13, Channels 1, 2, 3, and 4 are, respectively, the reference signal r, the VCM control signal u_v, the microactuator control signal $sat(u_m)$, and the displacement y. The seeking time is 0.1 ms and no overshoot is observed. The microactuator control signal is just about

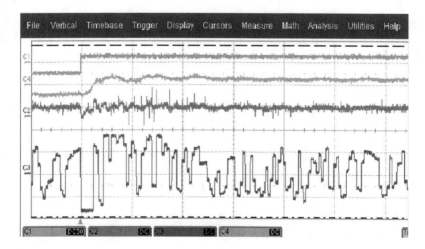

FIGURE 3.13
40 nm step response of the dual-stage system with the MTA. Timebase: 500 μs/div; C1 (500 mV/div): reference; C2 (50 mV/div): VCM control signal u_v; C3(2 V/div): MTA control signal $sat(u_m)$; C4 (500 mV/div): total displacement y. (From Ref. [4].)

to saturate. Moreover, Figure 3.14 shows the estimated microactuator output $\hat{y}_m$ and the corresponding microactuator control signal $sat(u_m)$ during a 40 nm step tracking.

Figure 3.15 shows that the thermal microactuator is able to move 15 nm, given by a 4 kHz sinusoid signal used as the reference. Channels 1, 2, 3, and 4 are, respectively, the reference signal r, the estimated microactuator output $\hat{y}_m$, the microactuator control signal $sat(u_m)$, and the displacement y.

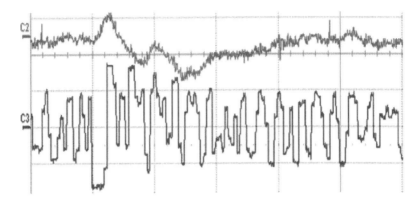

FIGURE 3.14
C2(200 mV/div): estimated MTA output $\hat{y}_m$; C3(2 V/div): MTA control signal $sat(u_m)$. (From Ref. [4].)

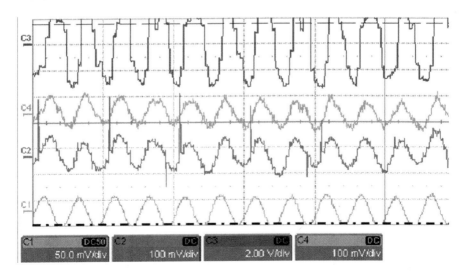

FIGURE 3.15
Signals of the dual-stage system with a 4 kHz sinusoid signal as the reference. Timebase: 500 µs/div; C1 (50 mV/div): reference; C2 (100 mV/div): $\hat{y}_m$; C3(2V/div): MTA control signal $sat(u_m)$; C4(100 mV/div): total displacement y. (From Ref. [4].)

Dual-Stage Actuation Systems

3.5 Conclusion

The control design and implementation for the dual-stage actuation system with the thermal microactuator has been reported in this chapter. The decoupled control structure has been adopted for the control of this thermal microactuator based dual-stage actuation system. Particularly, the driving method and the modeling of the thermal microactuator have been discussed in detail, which is important to make effective use of the microactuator. The dual-stage closed-loop control system has demonstrated 6 kHz bandwidth with 6.9 dB and 44° stability margins. Such a high bandwidth is potentially able to support ultra-high-precision positioning in nanometer level.

References

1. J. Yang, G. K. Lau, C. P. Tan, N. B. Chong, B. Thubthimthong, and Z. He, An electro-thermal micro-actuator based on polymer composite for application to dual-stage positioning systems of hard disk drives, *Sens. Actuat. A Phys.*, 187, pp. 98–104, 2012.
2. J. Yang, G. K. Lau, C. P. Tan, N. B. Chong, B. Thubthimthong, L. Gonzaga, and Z. He, Silicon-polymer composite electro-thermal microactuator for high track density HDD, *ASME-ISPS/JSME-IIP Joint International Conf. on Micromechatronics for Information and Precision Equipment*, Santa Clara University, CA, US, June 18–20, 2012, pp. 66–68.
3. G. K. Lau, J. Yang, B. Thubthimthong, N. B. Chong, C. P. Tan, and Z. He, Fast electrothermally activated micro-positioner using a high-aspect-ratio micromachined polymeric composite, *Appl. Phys. Lett.*, 101, p. 033108, 2012.
4. T. Gao, C. Du, C. P. Tan, Z. He, J. Yang, and L. Xie, High bandwidth control design and implementation for a dual-stage actuation system with a micro thermal actuator, *IEEE Trans. Magn.*, 49(3), pp. 1082–1087, 2013.
5. G. Guo, D. Wu, and T. C. Chong, Modified dual-stage controller for dealing with secondary-stage actuator saturation, *IEEE Trans. Magn.*, 39(6), pp. 3587–3592, 2003.

4

Modeling and Control of a Three-Stage Actuation System

4.1 Introduction

As stated in Chapter 1, in view of the additional bandwidth requirement which is limited by stroke constraint and saturation of secondary actuators, three-stage actuation systems are thereby proposed to meet the demand of a higher bandwidth. In this chapter, a specific three-stage actuation system is presented and a control strategy is proposed for the three-stage actuation system, which is based on a decoupled master–slave dual-stage control structure [1] combined with a third-stage actuation in parallel format [2]. This control strategy makes it easy to push the servo bandwidth and meet the performance requirement of the overall control loop. Nevertheless, other control configurations for the three-stage actuation systems are also presented and discussed in this chapter.

4.2 Actuator and Vibration Modeling

In the three-stage actuation system having the structure in Figure 1.8, VCM (voice coil motor) actuator as the first-stage actuator works as the primary actuator, and the PZT (Pb-Zr-Ti) milliactuator as the second-stage actuator is used to strengthen the actuation system with a reasonable bandwidth. A third-stage actuator which is in a slider- or head-level is then needed to further push the bandwidth. The thermal microactuator previously discussed in Chapter 3 is used as the third-stage actuator.

As shown in Figure 1.8, the VCM actuator is denoted by $P_v(s)$, the PZT milliactuator is denoted by $P_p(s)$, and the thermal microactuator is denoted by $P_m(s)$. Figures 4.1–4.3 show their measured frequency responses and the modeled ones. The VCM actuator and the PZT milliactuator have the shared first dominant resonance at 5 kHz. The PZT milliactuator has 200 nm stroke with input constraint ±10 V, and the thermal microactuator has 50 nm stroke with input constraint ±5 V.

47

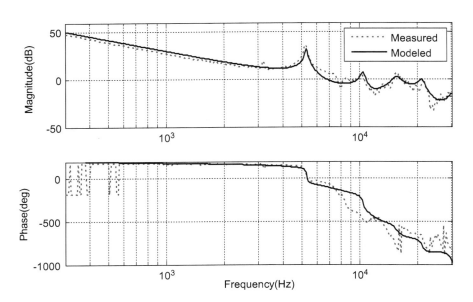

FIGURE 4.1
Frequency responses of the VCM actuator as the first stage. (From Ref. [2].)

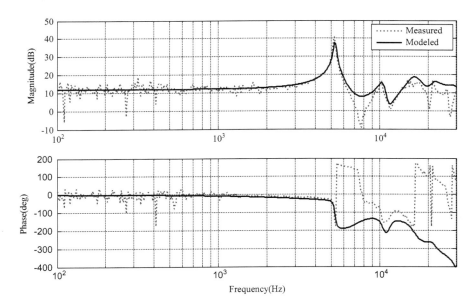

FIGURE 4.2
Frequency responses of the PZT milliactuator as the second stage. (From Ref. [2].)

Three-Stage Actuation System

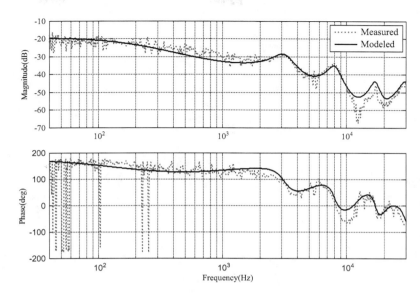

FIGURE 4.3
Frequency responses of the thermal microactuator as the third stage. (From Ref. [2].)

Remark 4.1

In hard disk drives, a large skew angle will affect the slider's flying performance and offtrack capability and lead to an increase in side reading and an offset of written transitions from the track center. This complicates the position error signal calibration process for the servo loop, especially for the recording technology [3–8] where small skew is even more necessary. An effective way to achieve a small skew angle change is to extend the arm length of the VCM actuator [9,10]. Figure 4.4 is used to illustratively show the skew angle range with an arm length. However, the VCM actuator will become weaker and will not be capable of supporting a higher bandwidth, as its resonant frequency is lower and there could be more uncertainty in the low frequency range. The introduced three-stage actuation system could be one solution to meet the small skew angle requirement in addition to the additional bandwidth requirement, which is limited by the secondary actuator's stroke constraint and saturation.

In hard disk drives, there are various disturbances mainly due to torque disturbances from spindle motor, actuator pivot friction, airflow induced non-repeatable disk, suspension and slider vibrations, and mechanical resonance vibration [11]. These are internal disturbances inside the hard disk drive. Let d_1 and d_2 denote input and output disturbances, respectively. A spectrum of the system disturbance ($P_v d_1 + d_2$) is shown in Figure 4.5, where the frequency content concentrates on 0–8 kHz and the high frequency disturbance around 10 kHz and 15 kHz are simulated by considering the mechanical resonances of the actuation system excited by air turbulence.

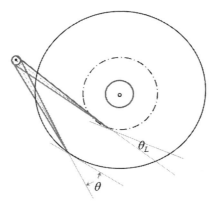

$\theta \in [-16°,+16°] \rightarrow \theta_L \in [-4°,+4°]$ with longer actuator arm

FIGURE 4.4
An illustrative diagram to show the skew angle (θ and θ_L) range with arm length. (From Ref. [2].)

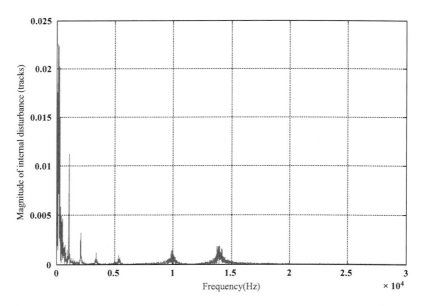

FIGURE 4.5
Internal disturbance spectrum. (From Ref. [2].)

In addition to the internal disturbances, external vibrations also have to be considered when designing the servo control loop. A typical external vibration in HDD industrial companies' interest is the vibe vibration coming from the speakers in computers [12]. The spectrum of an external vibration is shown in Figure 4.6, where, in addition to the large amplitude in low frequency range, there is an obvious spectrum from 1 to 3 kHz that is due to the

Three-Stage Actuation System

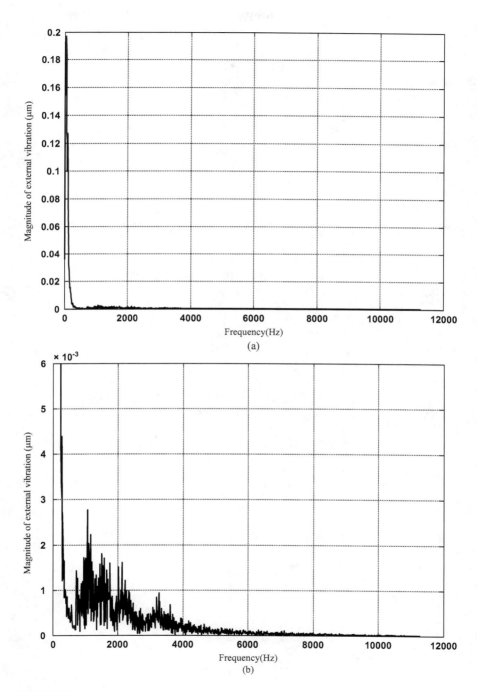

FIGURE 4.6
(a) Spectrum of external vibration d. (b) Zoomed-in view of (a). (From Ref. [2].)

vibe vibration. The low frequency part can be attenuated sufficiently by the closed-loop servo system, while the high frequency part is difficult to reduce due to a limited servo bandwidth.

Because of the external vibration, which is usually much higher than the internal disturbance, the strokes of the second and the third actuators must be high enough so that there will not be saturation in the closed-loop system and the control system performance will not deteriorate. In another word, higher stroke actuators have stronger resistance abilities to external vibration and ensure that the whole control system can keep a desired control performance.

4.3 Control Strategy and Controller Design

Figure 4.7 shows the control structure for the three-stage actuation system, where n stands for the measurement noise of the position error signal and the reference $r = 0$, in addition to the internal disturbances d_1 and d_2 and the external vibration d. The control scheme is based on the decoupled master–slave dual-stage control [1] and the third-stage microactuator is added in parallel with the dual-stage control system. The parallel format is advantageous to the overall control bandwidth enhancement, especially for the microactuator having limited stroke which restricts the bandwidth of its own loop. The reason why the decoupled control structure is adopted here is that its overall sensitivity function is the product of those of the two individual loops, and the VCM actuator's and the PZT milliactuator's controllers can be designed separately, which simplifies the overall control system design.

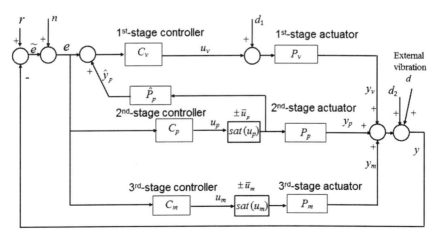

FIGURE 4.7
Control system for the three-stage actuation system. (From Ref. [2].)

Three-Stage Actuation System   53

In Figure 4.7, $\hat{P}_p$ is the approximated model of P_p for the decoupled dual-stage control structure, C_v, C_p, and C_m are the controllers to be designed for each actuator. $\bar{u}_p$ and $\bar{u}_m$ are, respectively, the saturation levels of the second- and the third-stage actuators.

The open-loop transfer functions for the individual actuator loops are respectively written as:

$$G_v(z) = P_v(z)C_v(z) \tag{4.1}$$

$$G_p(z) = P_p(z)C_p(z) \tag{4.2}$$

and

$$G_m(z) = P_m(z)C_m(z). \tag{4.3}$$

The open-loop transfer function and the sensitivity function of the dual-stage actuation system are respectively given by

$$G_{\text{dual}}(z) = G_v(z) + G_p(z) + G_v(z)G_p(z) \tag{4.4}$$

and

$$S_{\text{dual}}(z) = \frac{1}{1 + G_{\text{dual}}(z)}. \tag{4.5}$$

The open-loop transfer function of the three-stage actuation system is derived as

$$G(z) = G_v(z) + G_p(z) + G_v(z)G_p(z) + G_m(z) \tag{4.6}$$

and the overall sensitivity function is given by

$$S(z) = 1/\left[1 + G(z)\right]. \tag{4.7}$$

The specifications for the control design of the three-stage actuation system are prescribed as follows:

The VCM actuator $P_v(s)$, as the first-stage actuator, works in a low bandwidth below 1 kHz. The PZT actuated milliactuator $P_p(s)$ used as the second-stage actuator works under a reasonably high bandwidth up to 3 kHz. The third-stage actuator $P_m(s)$ which is a thermal microactuator is used to further push the bandwidth as high as possible. As discussed in Chapter 3, the thermal microactuator is driven by power to deal with its nonlinearity issue.

The design of each controller uses the loop shaping method stated in Chapter 2 as the method is conveniently and effectively applicable to obtain appropriate controllers by adjusting the weighting function parameters according to the control performance requirements.

The sampling rate of the control system is 60 kHz. The VCM control loop is designed having a bandwidth of 600 Hz, as seen in Figure 4.8. The control performances of both the VCM actuator and PZT milliactuator are limited by their dominant resonance modes. The controller of the second-stage PZT milliactuator is designed to let the open loop have a sufficiently high gain in the low frequency range and a high bandwidth as well. The third-stage microactuator is mainly used to push the bandwidth further. The open-loop frequency responses of the VCM actuator, the second-stage milliactuator and the third-stage microactuator loops, i.e., $G_v(z)$, $G_p(z)$, and $G_m(z)$ are all shown in Figure 4.8. The third-stage microactuator loop has the highest bandwidth and a lower gain in a low frequency than the second-stage milliactuator. Due to more limited stroke and constraint input, the third-stage microactuator loop has to be designed to have a smaller gain in the low frequency range for a higher bandwidth.

As shown in Figure 4.9, the dual-stage bandwidth is 2.3 kHz, and the overall three-stage system achieves more than 6 kHz bandwidth. Note that 6 kHz bandwidth is not possibly achieved by the dual-stage system due to the 5 kHz dominant resonance of the PZT milliactuator. The overall sensitivity function $S(z)$ is shown in Figure 4.10, which means that the three-stage system possesses the vibration rejection capability in a broad frequency range up to 6 kHz.

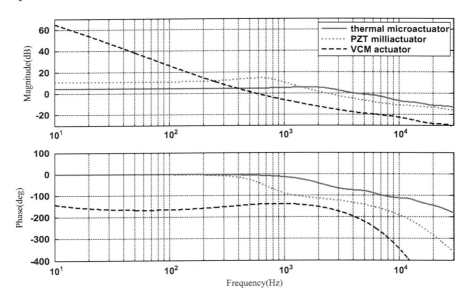

FIGURE 4.8
Frequency responses of the open-loop transfer functions (Solid line: $G_m(z)$. Dotted line: $G_p(z)$. Dashed line: $G_v(z)$.) (From Ref. [2].)

Three-Stage Actuation System

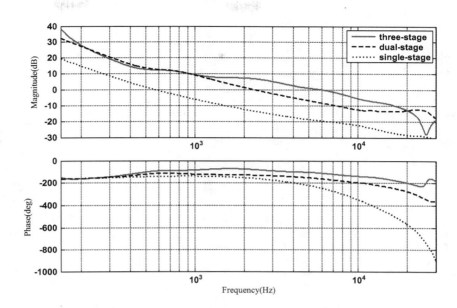

FIGURE 4.9
Frequency responses of the VCM single-stage, the dual-stage, and the three-stage open-loop transfer functions, i.e., $G_v(z)$, $G_{dual}(z)$, and $G(z)$. (6.1 kHz servo bandwidth (defined as 0 dB crossover frequency)) (Solid line: three-stage; Dotted line: single-stage; Dashed line: dual-stage.) (From Ref. [2].)

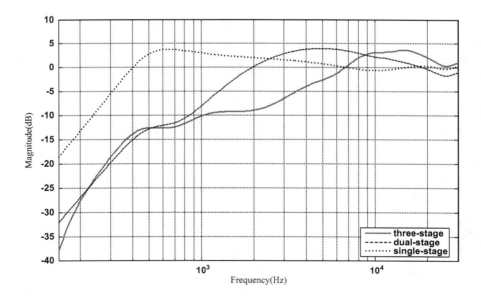

FIGURE 4.10
Sensitivity functions of the VCM single-stage (dotted line), the dual-stage (dashed line), and the three-stage (solid line) loops, i.e., $S_v(z)$, $S_{dual}(z)$, and $S(z)$. (From Ref. [2].)

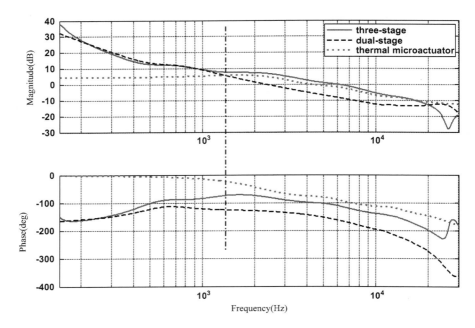

FIGURE 4.11
Frequency responses of the dual-stage (dashed line), the thermal microactuator (dotted line), and the three-stage (solid line) open-loop transfer functions, i.e., $G_{dual}(z)$, $G_m(z)$, and $G(z)$, indicating the hand off frequency and phase from the dual-stage to the three-stage. (From Ref. [2].)

As indicated in Figure 4.11, a smaller phase difference at the hand off frequency between the third-stage microactuator and the dual-stage implies a better error rejection performance and it is suggested to be within 100°.

4.4 Performance Evaluation

In the presence of internal disturbances d_1, d_2 and measurement noise n, the error signal $\tilde{e}$ equals to

$$\tilde{e} = -y = -S(P_v d_1 + d_2) - (1-S)n \tag{4.8}$$

based on which a track density without the consideration of external vibration is determined. Here, 3σ of measurement noise n is assumed to be 2% of track width, and d_1 and d_2 are scaled so that the track density of 800k tracks/inch is achieved.

With the three-stage control, the spectrum of the error signal $\tilde{e}$ is shown in Figure 4.12, and the internal disturbance is reduced by 80% in the position

Three-Stage Actuation System

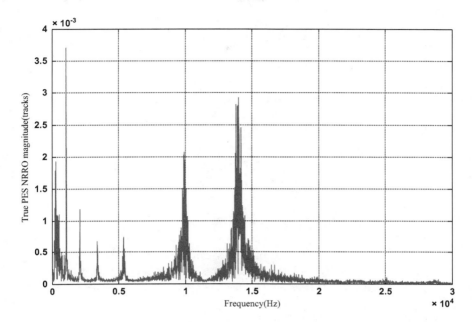

FIGURE 4.12
Spectrum of the error signal $\tilde{e}$ of the three-stage system under control. (From Ref. [2].)

error signal. The spikes noticed in Figure 4.12 could be suppressed further by a higher bandwidth and accordingly more freedom is available to shape the control loop. The apparent high peaks near 10 kHz and 15 kHz in Figure 4.12, visible in Figure 4.5 as well, are due to the hump above 0 dB of the sensitivity function as seen in Figure 4.10. By the way, an application of sensing technology for the excited mechanical vibration is one choice to attenuate these peaks by active vibration control method.

External vibration from the system working environment is much higher than the internal disturbance, especially for ultra-high precision positioning systems. In the presence of external vibration, the actuators' control effort is dominantly determined by the external vibration. But because the actuator input is constrained, the external vibration level has to be limited. Otherwise, saturation will occur in the control loop and the control system performance will be degraded. Therefore, the stroke specification of the actuators, especially milliactuators and microactuators, is very important for achievable control performance. Higher stroke actuators have stronger abilities to make sure that the control performances are not degraded in the presence of external vibrations.

The control efforts of the PZT milliactuator and the microactuator are subsequently investigated under certain levels of external vibration d as well as the previous internal disturbances and noise.

With the internal disturbance d_1, d_2, the measurement noise n, and the external vibration d, the position error e is determined as

$$e = -S(P_v d_1 + d_2 + d_e) + Sn \qquad (4.9)$$

The control signals and the positions of the PZT milliactuator, the microactuator, and the VCM actuator are given by

$$u_p = C_p e, y_p = P_p C_p e \qquad (4.10)$$

$$u_m = C_m e, y_m = P_m C_m e \qquad (4.11)$$

$$u_v = C_v \left(1 + \hat{P}_p C_p\right) e, y_v = P_v \left(u_v + d_1\right) \qquad (4.12)$$

Table 4.1 shows the calculated values for four cases, where there is no saturation in the overall control system. In another word, the control system works without saturation and possesses the control performance as a linear system.

The first case tells that to let the third-stage microactuator's control signal be within ±5 V, the external vibration cannot be higher than 0.43 µm. The second case means that if the microactuator's stroke is doubled to be 100 nm, then a higher level of the external vibration, i.e., 1.2 µm, is supported.

In the third case, assuming that the microactuator's stroke is much increased, we redesign the controller to have a higher control bandwidth, say 7.2 kHz, as seen in Figure 4.13. In this situation, the external vibration has to be lower so that the microactuator's control signal is within ±5 V.

The fourth case assumes that the VCM actuator becomes weaker. Figure 4.14 shows that the PZT milliactuator loop is designed to have a higher gain and cover more displacement, and with the same third-stage microactuator loop, the same overall bandwidth of 6.1 kHz is achieved. In this case, the PZT milliactuator requires more control effort (2.6 V), and about the same level of external vibration (0.4 µm) is supported as in the first case.

The analysis results listed in Table 4.1 imply that the controller design for the microactuators with input constraint must take into account both external vibration requirement and actuators' stroke, based on which an appropriate bandwidth should be decided when designing the control system.

TABLE 4.1

Bandwidth and Control Signal Corresponding to Stroke and External Vibration Level

External vibration d (3σ (µm))	0.43	1.2	1.0	0.4
Third-stage microactuator stroke (nm)	50	100	100	50
Bandwidth (kHz)	6.1	6.1	7.2	6.1
Second-stage milliactuator control signal (3σ (v))	1	2.5	2.2	2.6

Three-Stage Actuation System 59

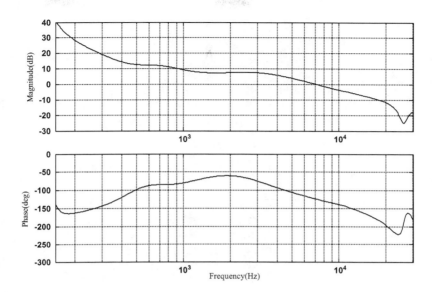

FIGURE 4.13
Open-loop frequency responses of the three-stage actuation system under control (7.2 kHz bandwidth). (From Ref. [2].)

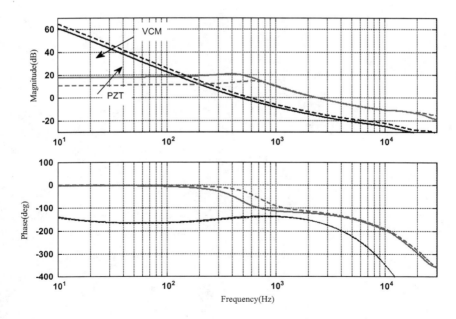

FIGURE 4.14
Open-loop frequency responses of VCM actuator and PZT milliactuator loops (arrows indicate that VCM becomes weaker and PZT becomes stronger with comparison to those in Figure 4.8). (From Ref. [2].)

It should be noted that the supported external vibration level is obtained corresponding to the profiles in Figures 4.5 and 4.6. For different profiles, the external vibration level will be different very probably. Therefore, the required stroke still depends on the frequency content of the vibration, especially the external vibration, in addition to its amplitude. In this chapter, the specific external vibration having the spectrum in Figure 4.6 is typically used, and the analysis methodology applied here applies to any vibration.

4.5 Experimental Results

Experiment was conducted to verify the implementability and the control performance of the closed-loop control of the three-stage system. As usual, it uses a LDV (laser Doppler vibrometer) to measure the displacement y, and a dSPACE 1103 to implement the controllers.

The experimental and simulated frequency responses of $G(z)$ for the three-stage closed-loop system are plotted in Figure 4.15, which shows that the bandwidth is 6 kHz. It is noticed that the magnitude is close to 0 dB at frequencies near and higher than the bandwidth, in order to avoid saturation, since raising the loop gain needs more stroke.

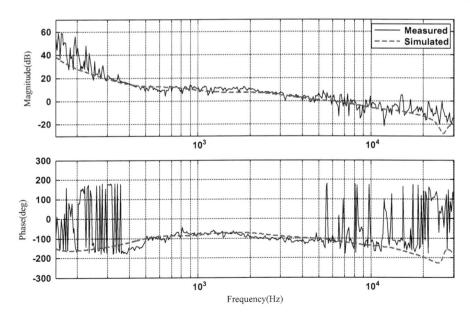

FIGURE 4.15
Open-loop $G(z)$ frequency responses of the three-stage actuation system with controllers (experimental and simulation results). (From Ref. [2].)

Three-Stage Actuation System

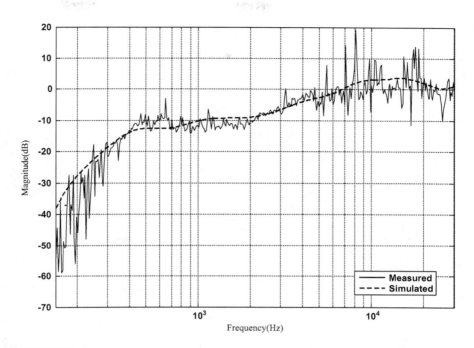

FIGURE 4.16
Magnitude of the sensitivity function S(z) (experimental (solid line) and simulation (dashed line) results). (From Ref. [2].)

The magnitude of the sensitivity function $S(z)$ of the three-stage closed-loop control system is shown in Figure 4.16, where it is observed that the error rejection capability is up to 6 kHz.

Step response performance is also tested for the closed loop, and the step response with reference $r = 25$ nm is shown in Figure 4.17. The time taken to reach the target is 0.1 ms. The PZT milliactuator control signal, the thermal microactuator control signal $sat(u_m)$, and the position y are displayed in channels 1, 3, and 4, respectively.

4.6 Different Configurations of the Control System

In Figure 4.18, the control for the third-stage microactuator P_m involves its approximated model $\hat{P}_m$, which generates an estimated displacement $\hat{y}_m$ added into the position error signal e. The overall sensitivity function,

$$S(z) \approx S_v(z)S_p(z)S_m(z) \tag{4.13}$$

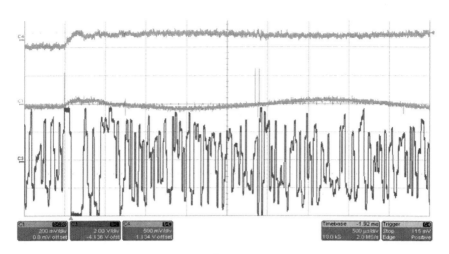

FIGURE 4.17
25 nm step response of the three-stage control system (Channel 4: position y; Channel 1: PZT milliactuator control signal; Channel 3: thermal microactuator control signal). (seeking time: 0.1 ms). (From Ref. [2].)

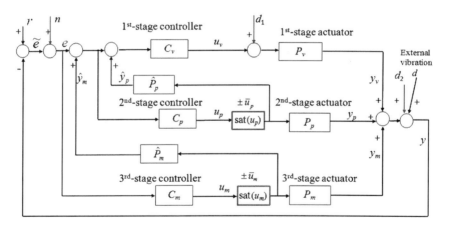

FIGURE 4.18
Another control structure for the three-stage actuation system with the third-stage decoupled.

where $S_v(z)$ and $S_p(z)$ are in Equation (2.4), and

$$S_m(z) = \frac{1}{1 + P_m(z)C_m(z)}. \tag{4.14}$$

It means that the sensitivity function of the three-stage actuation system is the product of the three individual sensitivity functions: $S_v(z)$, $S_p(z)$, $S_m(z)$. Denote the dual-stage open-loop transfer function as G_d, which is given by

$$G_d(z) = G_v(z) + G_p(z) + G_v(z)G_p(z). \quad (4.15)$$

The open-loop transfer function of the overall system is

$$G(z) = G_d(z) + G_m(z) + G_d(z)G_m(z). \quad (4.16)$$

The control signals and the positions of the PZT milliactuator, the microactuator, and the VCM actuator are given by

$$u_p = C_p\left(1 + \hat{P}_m C_m\right)e, y_p = P_p u_p \quad (4.17)$$

$$u_m = C_m e, y_m = P_m C_m e \quad (4.18)$$

$$u_v = C_v\left(1 + \hat{P}_p C_p\right)\left(1 + \hat{P}_m C_m\right)e, y_v = P_v u_v. \quad (4.19)$$

For simplicity, controllers $C_v(z)$ and $C_p(z)$ are designed similarly as those in Section 4.3. For the control structure in Figure 4.18, to have the same bandwidth as that in Section 4.3, the open-loop transfer function $P_m(z)C_m(z)$ of the third-stage microactuator needs to have a higher bandwidth. As seen in Figure 4.19, the bandwidth is increased from 4.2 kHz to 6.4 kHz. Figures 4.20 and 4.21 show the comparison of the open-loop transfer functions and the

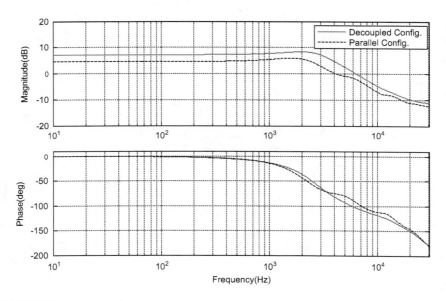

FIGURE 4.19
Frequency responses of $P_m(z)C_m(z)$ of the third-stage microactuator configured by parallel and decoupled structures.

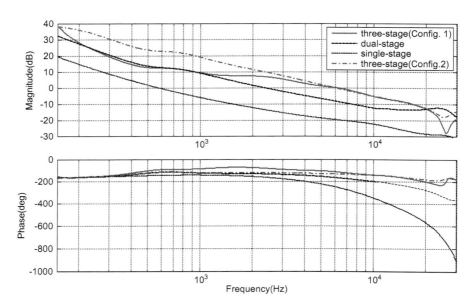

FIGURE 4.20
Frequency responses of the open-loop transfer functions $G(z)$ in Equations (4.6) and (4.16) of the control structures in Figures 4.7 (Config. 1) and 4.18 (Config. 2), respectively.

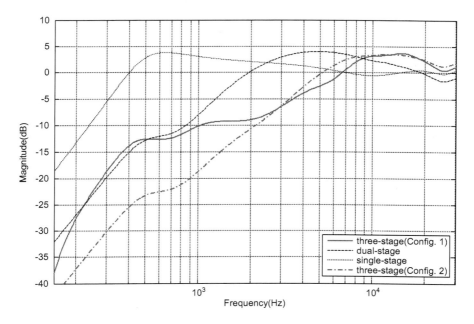

FIGURE 4.21
Sensitivity function comparison of the control structures in Figures 4.7 (Config. 1) and 4.18 (Config. 2).

sensitivity functions between the two control configurations in Figures 4.7 and 4.18. The decoupled configuration in Figure 4.18 makes the low frequency gain much higher, and consequently there is much better rejection capability within 2 kHz, as observed in Figures 4.20 and 4.21, respectively, compared to the parallel configuration in Figure 4.7. The disturbance rejection difference is also reflected in the spectrum of the error signal $\tilde{e}$ shown in Figure 4.22, compared with that in Figure 4.12.

Table 4.2 gives further comparison between the two configurations in Figures 4.7 and 4.18. Because of the increased bandwidth of the third-stage microactuator loop, much less external vibration is supported in order to make the microactuator input u_m within ±5 V and avoid saturation. The second-stage milliactuator control signal u_p is increased due to the added-on signal $\hat{y}_m$ for decoupling, which is also seen in Equation (4.17).

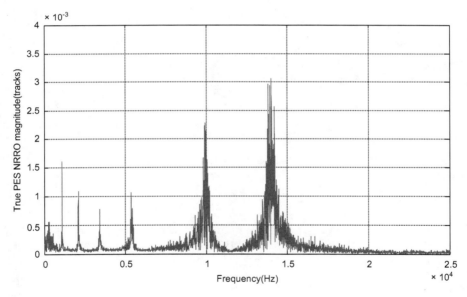

FIGURE 4.22
Spectrum of the error signal $\tilde{e}$ in Figure 4.18.

TABLE 4.2

Gain Margin, Phase Margin, Control Signal Corresponding to the External Vibration (Bandwidth: 6.1 kHz)

Control Structure	Figure 4.7	Figure 4.18
External vibration d (3σ (μm))	0.43	0.2
Second-stage milliactuator control signal (3σ (v))	0.9	1.5
Gain margin (dB)	11	13
Phase margin (°)	68	51

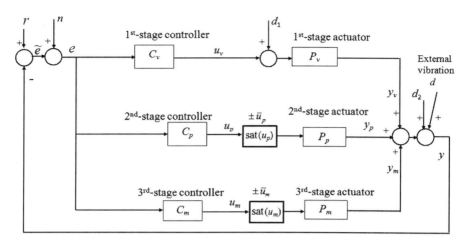

FIGURE 4.23
Parallel control structure for the three-stage actuation systems.

In addition, Figure 4.23 gives the parallel control structure for the three actuators in three-stage actuation systems. This control structure has been adopted in [13], where there are details about the design of the controllers.

4.7 Conclusion

The three-stage actuation system involving the VCM primary actuator, PZT actuated milliactuator, and thermal microactuator has been proposed to enable extremely high control bandwidth. The VCM actuator and PZT milliactuator used in this chapter have the first dominant resonance at 5 kHz, the thermal microactuator stroke is 50 nm with ±5 V input limitation, and the PZT milliactuator stroke is 200 nm, much higher than that of the microactuator.

For the designed control system, the PZT milliactuator provides a reasonable bandwidth of 2.3 kHz and could be used to strengthen the system for the weak VCM primary actuator. The microactuator has been used to push the bandwidth to more than 6 kHz. The relationship among the external vibration, the microactuator stroke, and the achievable control bandwidth has been discussed for being considered in the controller design. For the certain external vibration profile considered in this chapter, the external vibration level has been determined so that the control system can support it. The discussion suggests that in addition to the traditional wisdom of just increasing the resonant frequency, adding more stroke to the microactuator will give more freedom to the loop shaping for the control system design.

References

1. K. Mori, T. Munemoto, H. Otsuki, Y. Yamaguchi, and K. Akagi, A dual-stage magnetic disk drive actuator using a piezoelectric device for a high track density, *IEEE Trans. Magn.*, 27(6), pp. 5298–5300, 1991.
2. C. Du, C. P. Tan, and J. Yang, Three-stage control for high servo bandwidth and small skew actuation, *IEEE Trans. Magn.*, 51(1), p. 310017, 2015.
3. J. Van Ek, A. Shukh, E. Murdock, G. Parker, and S. Batra, Micromagnetic perpendicular recording model: Soft magnetic underlayer and skew effect, *J. Magn. Magn. Mater.*, 235, pp. 408–412, 2001.
4. R. Wood, Y. Sonobe, Z. Jin, and B. Wilson, Perpendicular recording: The promise and the problems, *J. Magn. Magn. Mater.*, 235, pp. 1–9, 2001.
5. M. H. Kryder, E. C. Gage, T. W. McDaniel, W. A. Challener, R. E. Rottmayer, G. Ju, Y. T. Hsia, and M. F. Erden, Heat assisted magnetic recording, *Proc. IEEE*, 96(11), pp. 1810–1835, 2008.
6. B. X. Xu, Z. H. Cen, J. H. Goh, J. M. Li, Y. T. Toh, J. Zhang, K. D. Ye, and C. G. Quan, Performance benefits from pulsed laser heating in heat assisted magnetic recording, *J. Appl. Phys.*, 115, pp. 17B701-1–17B701-3, 2014.
7. M. F. Erden and J. Gadbois, Two dimensional magnetic sensor immune to skew angle misalignment, US patent No. 0286502, 31 Oct. 2013.
8. R. Wood, M. Williams, A. Kavcic, and J. Miles, The feasibility of magnetic recording at 10 Terabits per square inch on conventional media, *IEEE. Trans. Magn.*, 45(2), pp. 917–923, 2009.
9. Z. He, E. H. Ong, and G. Guo, Optimization of a magnetic disk drive actuator with small skew actuation, *J. Appl. Phys.*, 91(10), pp. 8709–8711, 2002.
10. J. Akiyama, T. Inoue, K. Higashi, Y. Ohtsubo, and K. Tanimoto, Magnetic disk drive having a constant skew angle, US patent No.6021024, 2000.
11. C. Du and L. Xie, *Modeling and Control of Vibration in Mechanical Systems*, CRC press, Boca Raton, FL, 2010.
12. "HDD and speaker mounting optimization for vibration reduction in notebook computers", 2012, http://wenku.baidu.com/view/8055956ea45177232f60a283.html.
13. T. Atsumi, S. Nakamura, M. Furukawa, I. Naniwa, and J. Xu, Triple-stage-actuator system of head-positioning control in hard disk drives, *IEEE Trans. Magn.*, 49(6), pp. 2738–2743, 2013.

5

Dual-Stage System Control Considering Secondary Actuator Stroke Limitation

5.1 Introduction

It is understood from the previous chapter that dual-stage system control is vital to the multi-stage system control. There are many common problems for dual-stage and multi-stage systems, such as microactuator stroke limitation or input constraint problems, saturation problem, hysteresis problem, etc. In this chapter as well as the subsequent several chapters, these specific problems will be focused for the dual-stage systems and appropriate solutions will be provided.

Properly designing base feedback controller is crucial and cost effective to fully utilizing the microactuator within limited stroke, although there are additional methods to solve the saturation issue such as the special decoupled master-slave control scheme in [1] and the anti-windup method [2]. Moreover, as technology advances, higher control sampling rate is available, the control system bandwidth can be pushed further [3] and theoretically more stroke is demanded for the microactuator to achieve a high control bandwidth. Therefore, it is necessary to take into account the stroke, the loop gain, and the tolerable vibration amplitude when designing the feedback controller, in order to avoid the saturation of the secondary actuator loop and ensure that it works in linear status and thus guarantee the achieved control performance.

Appropriate loop shaping is proposed in this chapter for the controller design of the secondary actuator. Specifically, high-gain, mid-gain, and low-gain loop shaping will be presented, and more freedom for the loop shaping is available by using a higher-order weighting function. In this chapter, a PZT (Pb-Zr-Ti) microactuator is used as the secondary actuator of the dual-stage actuation system.

69

5.2 More Freedom Loop Shaping for Microactuator Controller Design

Consider Figure 2.1, the widely applied dual-stage system control loop, where $\hat{P}_p(z)$ is represented by the static gain of the PZT microactuator and used to estimate the output of the PZT microactuator $P_p(s)$ for the dual-stage actuation decoupling purpose. The input limitation $\bar{u}_p$ is associated with the available stroke of the microactuator.

With the application of the control structure in Figure 2.1, the controllers $C_v(z)$ and $C_p(z)$ are designed independently. $C_v(z)$ is designed by using the H_∞ loop shaping method presented in Chapter 2 with the weighting function in (2.15). Particularly, as for the design of $C_p(z)$ for the microactuator, the weighting function form in "Remark 2.1" is used, where

$$W_c(s) = \frac{1}{M} \cdot \frac{s^2 + 2\xi_1 \omega s + \omega^2}{s^2 + 2\xi_2 \omega \sqrt{\varepsilon} s + \omega^2 \varepsilon} \tag{5.1}$$

and $W_m(s)$ is in (2.31). Figure 2.6 illustrates the weighting functions in the frequency domain, where the major difference of $\left|W^{-1}(s)\right|$ from $\left|W_c^{-1}(s)\right|$ is in the low-frequency range due to $\left|W_m^{-1}(s)\right|$. Therefore, it is understood that the loop shaping using the weighting function form in "Remark 2.1" gives more freedom to design a controller in order to meet more requirements. In what follows, the dual-stage system control having 5 kHz servo bandwidth will be designed using several sets of parameters for the weighting function to have different levels in the low-frequency range.

5.3 Dual-Stage System Control Design for 5 kHz Bandwidth

The PZT microactuator considered in this chapter is shown in Figure 5.1 having dominant resonance at 50 kHz.

The controllers $C_v(z)$ and $C_p(z)$ are designed to achieve 5 kHz control bandwidth with the sampling rate of 62.64 kHz. Figure 5.2 shows the frequency responses of $G_v(z)$, $G_p(z)$, and the dual-stage open-loop transfer function $G(z) = G_v(z) + G_p(z) + G_v(z)G_p(z)$. It is noticed that the VCM (voice coil motor) loop bandwidth is about 1.5 kHz and the PZT microactuator loop is nearly 5 kHz so that the overall dual-stage loop bandwidth can reach 5 kHz. 20 dB gain is noticed in the low-frequency range of $G_p(z)$.

The magnitudes of sensitivity functions, $S_v(z)$, $S_p(z)$, and $S(z)$, are plotted in Figure 5.3. By comparing $S(z)$ and $S_v(z)$, it is seen that the dual-stage control

Dual-Stage System Control

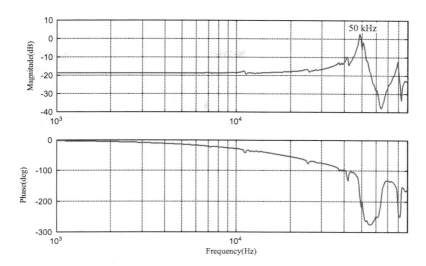

FIGURE 5.1
Frequency responses of a PZT microactuator $P_p(s)$.

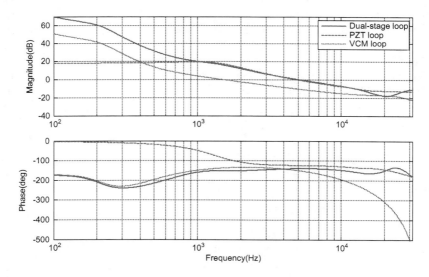

FIGURE 5.2
Open-loop frequency responses (5 kHz servo bandwidth) (VCM loop: $G_v(z)$; PZT loop: $G_p(z)$; and dual-stage loop: $G(z)$).

system has about 20 dB more attenuation capability to the vibrations within 1 kHz than the VCM single-stage loop, and it is able to reject the vibrations in the frequency range up to 3.5 kHz.

$W_c(s)$ in (5.1) is used for the controller $C_p(z)$ design, where $\varepsilon = 10^{-1}$, which implies the designed open-loop (i.e., $G_p(z)$) d.c. gain is about 20 dB, and the

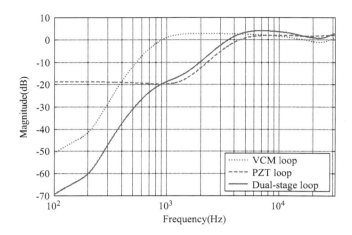

FIGURE 5.3
Sensitivity functions (VCM loop: $S_v(z)$; PZT loop: $S_p(z)$; and dual-stage loop: $S(z)$).

low-frequency magnitude of $S_p(z)$ is about −20 dB. These are observed in Figures 5.2 and 5.3. With $\varepsilon = 10^{-0.7}$, $C_p(z)$ is redesigned, the open-loop d.c. gain is lowered, and accordingly the low frequency magnitude of $S_p(z)$ is higher. The frequency responses of $G_p(z)$ attained from the above two designs are compared in Figure 5.4, namely high gain and mid-gain.

To further reduce the low frequency gain of $G_p(z)$, the weighting function in (2.30) is used. In $W_c(s)$, $\varepsilon = 10^{-0.8}$ and $\omega = 2\pi 4,470$. In $W_m(s)$, $\varepsilon_m = 10^{-1.2}$

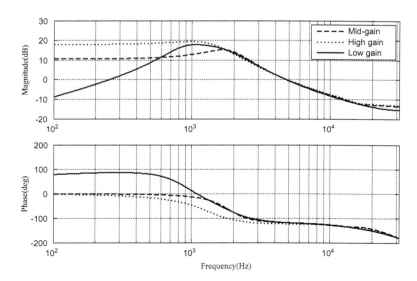

FIGURE 5.4
PZT open-loop $G_p(z)$ frequency responses.

Dual-Stage System Control

and $\omega_m = 2\pi 900$. The designed $G_p(z)$ with a small gain in the low-frequency range is shown in Figure 5.4 and compared with the previous two designs. Accordingly, the sensitivity functions $S_p(z)$, with the magnitude shown in Figure 5.5, have different levels in the low-frequency range. And, correspondingly, the overall dual-stage sensitivity functions $S(z)$ shown in Figure 5.6 have different attenuation levels for the vibrations in the frequency range of 1 kHz due to the designed different $S_p(z)$ or $C_p(z)$.

To this end, it should be noted that the weighting function in (2.33) associated with (2.34) and (2.15) or (5.1) enables us to have more freedom to shape

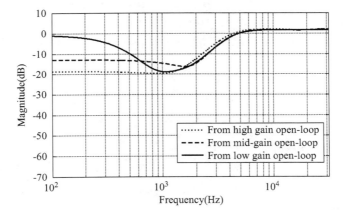

FIGURE 5.5
PZT loop sensitivity functions $S_p(z)$.

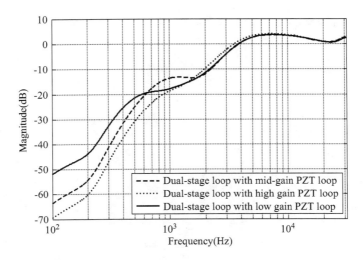

FIGURE 5.6
Dual-stage sensitivity functions $S(z)$.

the sensitivity function. Particularly, for the microactuators which have limited stroke, appropriate feedback controller needs to be designed to meet the stroke limitation by adjusting the weighting functions. Detailed evaluation and analysis is given in the next section to help understand the feedback controller design for limited stroke microactuators.

5.4 Evaluation with the Consideration of External Vibration and Microactuator Stroke

From Figure 2.1, the control signal and the displacement of the microactuator $P_p(s)$ denoted by u_p and y_p, respectively, are given by

$$u_p = C_p e, y_p = P_p C_p e \tag{5.2}$$

and e is in (2.10). And, 3σ value of noise n is assumed to be 1 nm.

The external vibration d_e focuses on 0–500 Hz, and its spectrum is seen in Figure 5.7. The microactuator stroke is determined according to the displacement or the control signal together with the microactuator's gain. As the external vibration level is higher, the error signal e becomes higher, leading to higher control signal u_p. The maximum external vibration level is thereby determined when the control signal $|u_p| \leq \bar{u}_p$, i.e., when the microactuator is not saturated [4]. Note that the microactuator loop is saturated

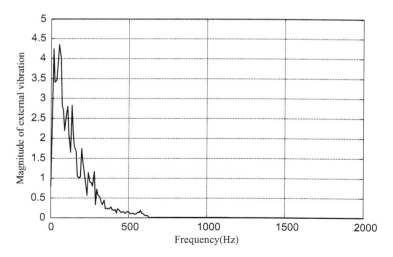

FIGURE 5.7
Power spectrum of vibration d_e (magnitude to be scaled).

means it doesn't work in the linear state and the control performance will be degraded.

With the presence of the external vibration d_e, the dual-stage system control aims to have the same 3σ value of the error signal $\tilde{e}$, say 5.7 nm, for the high-gain, mid-gain, and low-gain PZT loops. As seen in Table 5.1, the low-gain design requires small stroke length, but the maximum external vibration is less. Therefore, due to stroke limitation, the low-gain design is preferred, but the control loop cannot support a higher external vibration. If there exists a high external vibration in an environment where the control loop always needs to work linearly, more stroke is needed and a higher-gain design for the microactuator loop is then possible. As such, an appropriate design for the microactuator loop is desired, given by the availability of the microactuator stroke length and the external vibration conditions.

Without the consideration of the external vibration d_e, from (2.10), the errors e and $\tilde{e}$ are determined by

$$e = -S(P_v d_1 + d_2) + Sn \tag{5.3}$$

$$\tilde{e} = -S(P_v d_1 + d_2) - (1 - S)n. \tag{5.4}$$

The power spectrum of the error e is shown in Figure 5.8. With the high-gain PZT loop, the spectrum is much lower in the frequency range of 1 kHz. With

TABLE 5.1

Microactuator Stroke and Maximum External Vibration

PZT Loop Gain	High Gain	Mid-Gain	Low Gain
External vibration d_e (3σ (µm))	2.8	1.5	0.8
Stroke (nm)	±41	±19	±11
Error $\tilde{e}$ 3σ (nm)	5.7	5.7	5.7

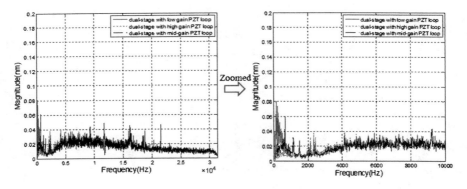

FIGURE 5.8
Power spectrum of the error e (no external vibration d_e).

the lower-gain PZT loops, the spectrum is higher, meaning less rejection of the disturbance in the range of 1 kHz. These results are attributed to the sensitivity functions in Figure 5.6. Furthermore, the 3σ values of the error $\tilde{e}$ are calculated. With the high-gain PZT loop, the 3σ value is 2.15 nm, and with the low-gain PZT loop, the error's 3σ is 2.2 nm in the dual-stage system control loop. The analysis implies that when the attenuation to the low-frequency disturbances is enough to some extent, it will not have a remarkable effect on the overall 3σ value.

It should be mentioned that although the secondary actuator is assumed to be a PZT microactuator in this chapter, the proposed methodology applies to other secondary actuators, such as milliactuator [5], and other types of microactuators [6].

5.5 Conclusion

For the controller design of dual-stage actuation systems, the secondary actuator loop having different loop gains in the low-frequency range has been designed by using the loop shaping method with higher-order weighting function for more freedom during the controller design. The PZT microactuator has been used as the secondary actuator in the dual-stage actuation system. As a result, the bandwidth of the designed dual-stage control system is as high as 5 kHz.

With the external vibration focusing on the frequency range of 0–500 Hz, the secondary actuator loop gain, required stroke, and allowable external vibration amplitude have been discussed. The analysis has shown that a proper controller design for the microactuator loop is desired, considering the availability of the microactuator stroke length and the external vibration conditions.

References

1. G. Guo, D. Wu, and T. C. Chong, Modified dual-stage controller for dealing with secondary-stage actuator saturation, *IEEE Trans. Magn.*, 49(6), pp. 3587–3592, 2003.
2. G. Herrmann, M. C. Turner, I. Postlethwaite, and G. Guo, Practical implementation of a novel anti-windup scheme in a HDD-dual-stage servo-system, *IEEE/ASME Trans. Mech.*, 9(3), pp. 580–592, 2004.
3. D. Abramovitch, T. Hurst, and D. Henze, The PES pareto method: uncovering the strata of position error signals in disk drives, *Proc. of the 1997 American Control Conference*, Albuquerque, NM, June 3–6, 1997, pp. 2888–2895.

4. C. Du, C. P. Tan, and J. Yang, Three-stage control for high servo bandwidth and small skew actuation, *IEEE Trans. Magn.*, 51(1), p. 3100107, 2015.
5. R. B. Evans, J. S. Griesbach, and W. C. Messner, Piezoelectric microactuator for dual stage control, *IEEE Trans. Magn.*, 35(2), pp. 977–982, 1999.
6. T. Hirano, M. White, H. Yang, K. Scott, S. Pattanaik, S. Arya, and F. Huang, A moving-slider MEMS actuator for high-bandwidth HDD tracking, *IEEE Trans. Magn.*, 40(4), pp. 3171–3173, 2004.

6

Saturation Control for Microactuators in Dual-Stage Actuation Systems

6.1 Introduction

Chapter 3 has presented the feedback controller design, simulation, and experimental testing results for the dual-stage actuation system with the thermal microactuator as the secondary actuator, and a 6 kHz servo bandwidth has been achieved. It is known that there is a saturation problem for microactuators, and the saturation will degrade the control system performance or even cause instability. The saturation occurs easily when the dual-stage actuators are doing long distance seeking or working under external vibrations such as large audio vibe. To help avoid the saturation, an enough stroke is thus required for microactuators. In this chapter, an anti-windup method is used to control the saturation as an additional compensation in the feedback loop. The anti-windup compensator will not affect the closed-loop system when there is no input saturation in the loop. In the presence of saturation in the loop, the anti-windup compensator can achieve good closed-loop performance as close as possible to the nominal linear control system, while the time period during which the control input is saturated is minimized. In this chapter, the anti-windup saturation compensator is designed via an LMI (linear matrix inequality) approach [1,2] based on the rigid part of the thermal microactuator only [3], and thus can be done separately from the basis feedback controller.

6.2 Modeling and Feedback Control

The modeling of the thermal microactuator is detailed in Section 3.2. The control structure of the dual-stage actuation system with the thermal microactuator is shown in Figure 6.1, where the decoupled master-slaver structure is adopted. $P_v(s)$ and $P_m(s)$, respectively, represent the VCM (voice coil motor)

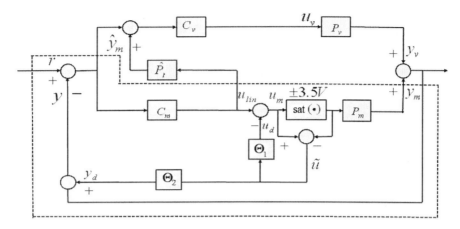

FIGURE 6.1
Dual-stage system control structure with anti-windup compensation. (From Ref. [3].)

actuator and the microactuator, $C_v(z)$ and $C_m(z)$ are their corresponding controllers, and $\hat{P}_t$ is an approximated model of the thermal microactuator. The details of the controller design and implementation are seen in Chapter 3. This chapter focuses on the saturation control of the thermal microactuator with the anti-windup compensation method. As seen in Figure 6.1, Θ_1 and Θ_2 are the anti-windup compensators to be designed for the saturation control.

6.3 Anti-Windup Compensation Design

The anti-windup compensation is realized by the two add-on compensators Θ_1 and Θ_2, as seen in Figure 6.1. Once the saturation occurs, the add-on compensators are enabled effectively to maintain stability and control performance. Without saturation occurrence, the whole system works as a nominal linear system. Having this anti-windup compensation, the system with the saturation approximates the nominal system as closely as possible and minimizes the impact of the saturation on the system nominal performance.

To design the anti-windup compensators, the anti-windup structure is decoupled into a nonlinear loop, a disturbance filter, and a nominal linear system, as shown in Figure 6.2. The dead zone function is defined as

$$D_z = u_m - \text{sat}(u_m) \tag{6.1}$$

with

$$u_m = u_{\text{lin}} - u_d. \tag{6.2}$$

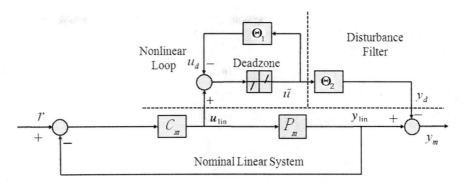

FIGURE 6.2
Equivalent representation of the anti-windup structure. (From Ref. [3].)

The nonlinear loop is to produce the signal u_d to cancel u_{lin} when the saturation occurs. The disturbance filter Θ_2 is used to reject the signal $\tilde{u}$, which is generated by the nonlinear loop.

The anti-windup compensators Θ_1 and Θ_2 are designed according to the following Theorem 6.1.

Theorem 6.1

There exist anti-windup compensators Θ_1 and Θ_2 that can achieve the performance $\|y_d\|_2 \leq \mu \|u_{lin}\|_2$ if there exists matrices $Q > 0$, $U > 0$, and L such that the LMI

$$\begin{bmatrix} -Q & * & * & * & * \\ -L & -2U & * & * & * \\ 0 & I & -\mu I & * & * \\ C_m Q^T + D_m L & D_m U^T & 0 & -I & * \\ A_m Q^T + B_m L & B_m U^T & 0 & 0 & -Q \end{bmatrix} < 0 \qquad (6.3)$$

holds, where * denotes an entry that can be deduced from the symmetric property of the matrix, and (A_m, B_m, C_m, D_m) is the state-space realization of $P_m(z)$. Then the anti-windup compensators Θ_1 and Θ_2 are given by $\Theta_1 : (A_m + B_m F, B_m, F, 0)$ and $\Theta_2 : (A_m + B_m F, B_m, C_m + D_m F, D_m)$, where

$$F = LQ^{-1}. \qquad (6.4)$$

This LMI-based full-order anti-windup compensation is easily implemented and has the fastest settling time and the shortest time needed to recover from saturation, compared with other anti-windup schemes such as IMC (internal model control)-type and suboptimal schemes [2].

6.4 Simulation and Experimental Results

The controller sampling rate used here is 60 kHz. Figure 6.3 shows the open-loop frequency responses of the VCM loop $G_v(z)$, the thermal microactuator (or MTA) loop $G_m(z)$, and the dual-stage actuation system $G(z)$ in (3.12). It is seen in Figure 6.3 that 6 kHz bandwidth, 5.13 dB gain margin, and 44° phase margin are achieved with the dual-stage actuation system. The sensitivity functions of the VCM loop $S_v(z)$, the microactuator loop $S_m(z)$, and the dual-stage actuation system $S(z)$ in (3.4)–(3.6) are also shown in Figure 6.4. It is noticed that the 0 dB frequency is 5 kHz and $|S(z)| = -15$ dB at 1 kHz, which means that in the dual-stage control system, the disturbances below 5 kHz are all rejected, and for that near the frequency of 1 kHz, the rejection capability can reach −15 dB.

To solve the saturation problem of the thermal microactuator in the dual-stage system, the anti-windup compensators Θ_1 and Θ_2 are designed just based on the rigid thermal part of the thermal microactuator. The simulation and the experiment have verified that the designed compensators are able to work effectively, which implies that the rigid part of the thermal microactuator plays the main role in the saturation.

To illustrate the effectiveness of the anti-windup compensation, a series of step responses are investigated for the dual-stage actuation system with the controls. The 20 nm step response comparison with and without the

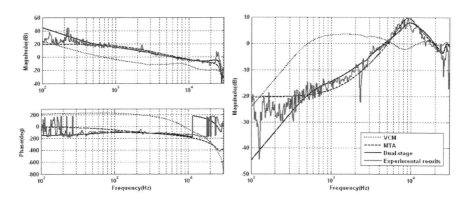

FIGURE 6.3
Open-loop frequency responses (left) and sensitivity functions (right) of the VCM loop, the microactuator (or MTA) loop, and the dual-stage system. (From Ref. [3].)

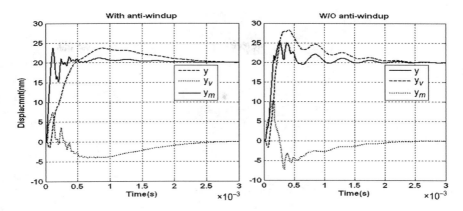

FIGURE 6.4
Displacement comparison with and without anti-windup compensation (20 nm step response). (From Ref. [3].)

anti-windup compensation is shown in Figure 6.4. The time taken to reach the target is 0.08 ms with the anti-windup compensation as compared to 0.15 ms without the anti-windup compensation. The larger overshoot and oscillation of the displacement y without the anti-windup compensation is mainly caused by the VCM output signal y_v. This is because during the period of the saturation (as seen in Figure 6.5) $|\hat{y}_m| > |y_m|$, and the effective VCM open loop will have a higher effective gain, which tends to drive the dual-stage system to become unstable. The corresponding control signals are plotted in Figure 6.5, which shows that the control signal escapes

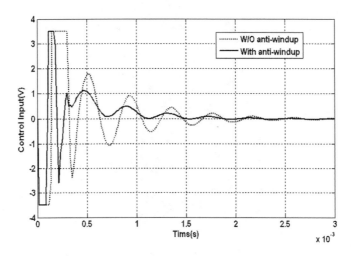

FIGURE 6.5
Comparison of the microactuator control signal (20 nm step response). (From Ref. [3].)

from saturation in 0.15 ms with the anti-windup compensation. In addition, with the microactuator gain variation of ±10%, the 20 nm step response performance does not change noticeably with the anti-windup compensation applied.

With the step length increased, the dual-stage system without the anti-windup compensation will become unstable. The anti-windup compensation is then enabled. To carry out the experiment with the anti-windup compensation, the experimental setup in Chapter 3 is used, where the displacement y is measured with an LDV (laser Doppler vibrometer) (range: 100 nm/V), and the controllers are implemented with dSPACE 1103.

Figure 6.6 shows the step response of 40 nm, and it is seen that the dual-stage actuation system reaches the target in 0.06 ms. The anti-windup compensators can help the dual-stage system to reach the target in a short time without increasing the overshoot. Also, Figure 6.7 shows the estimated microactuator output $\hat{y}_m$. Moreover, an 80 nm step response is shown in Figure 6.8. It takes 0.07 ms to reach the target while obvious oscillation appears. The oscillation could be caused by the mismatch of the plant model used in the anti-windup compensation structure.

In Chapter 3, a 40 nm step response has been shown with 0.1 ms seeking time achieved, where a signal limitation for $\hat{P}_t$ in the decoupling loop is used [4]. This helps maintain the whole system stability when saturated. However, on further increasing the seeking length, the output y_v of VCM will increase greatly. Meanwhile, the output y_m of the microactuator and the decoupled loop output $\hat{y}_m$ are limited by $sat(\cdot)$ function. Thus, as seen in the 60 nm step response

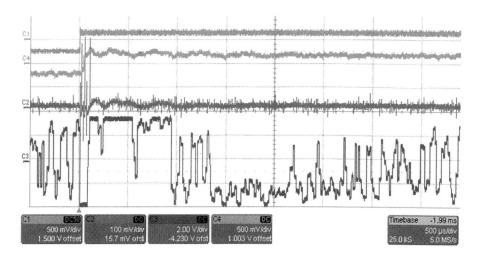

FIGURE 6.6
40 nm step response with anti-windup compensation (C1: reference; C2: VCM control signal u_v; C3: microactuator control signal $sat(u_m)$; C4: dual-stage displacement y). (From Ref. [3].)

Microactuator in Dual-Stage Actuation System

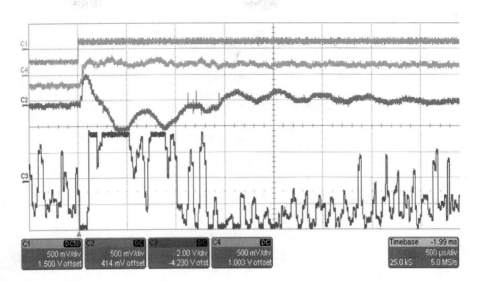

FIGURE 6.7
40 nm step response with anti-windup compensation (C1: reference; C2: estimated microactuator output $\hat{y}_m$; C3: microactuator control signal $sat(u_m)$; C4: dual-stage displacement y). (From Ref. [3].)

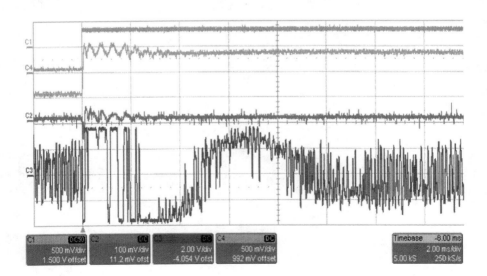

FIGURE 6.8
80 nm step response with anti-windup compensation (C1: reference; C2: VCM control signal u_v; C3: microactuator control signal $sat(u_m)$; C4: dual-stage displacement y). (From Ref. [3].)

in Figure 6.9, a large overshoot dominated by the VCM output y_v is observed, and the saturated control signal lasts for a longer time than that in Figure 6.8 where the anti-windup compensation is enabled to control the saturation.

Saturation will also degrade the performance of the closed-loop system in terms of vibration rejection. To evaluate the performance, the sensitivity function is calculated approximately with a swept sine wave injected to the closed loop when saturation occurs. The approximated sensitivity function is shown in Figure 6.10, with comparison to that of the linear system.

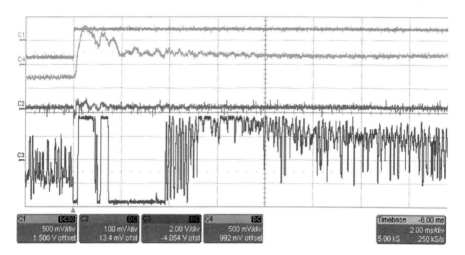

FIGURE 6.9
60 nm step response with signal limitation for $\hat{P}_t$ (C1: reference; C2: VCM control signal u_v; C3: microactuator control signal $sat(u_m)$; C4: dual-stage displacement y). (From Ref. [3].)

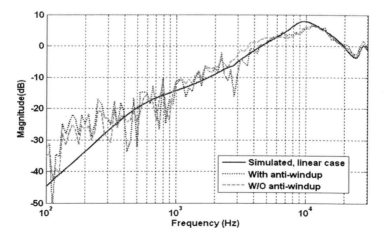

FIGURE 6.10
Sensitivity functions with saturation compared with the linear case. (From Ref. [3].)

Microactuator in Dual-Stage Actuation System

Because of the anti-windup compensation, the sensitivity function does not change much, while without the compensation it is affected by the saturation apparently.

The simulation and experimental results show that the decoupled master-slave structure is suitable to the thermal microactuator based dual-stage actuation system, while the electrostatic microactuator needs a capacitive sensor to measure its relative movement [5] if the decoupled master-slave control structure is adopted. Parallel control structure applies to both thermal microactuator and electrostatic MEMS (microelectromechanical systems) actuator as long as a sufficient stroke is available, whereas there is an effective stroke problem for the electrostatic MEMS actuator [6].

6.5 Conclusion

The anti-windup compensators have been designed and implemented for the thermal microactuator in the dual-stage actuation system. The design is based on the rigid part of the thermal microactuator model only, thus yielding low-order compensators. The low-order compensators have been implemented in the dual-stage actuation system, and it has been shown that the designed anti-windup compensators are able to work well in the presence of thermal microactuator saturation. Both simulation and experiment have shown that with the anti-windup compensation, the performance of the dual-stage actuation system has been improved much, and both step responses and vibration rejection capability have been investigated to show the effectiveness of the anti-windup compensation for the thermal microactuator.

References

1. J. M. Da Silva and S. Tarbouriech, Anti-windup design with guaranteed regions of stability for discrete-time linear systems, *Syst. Control Lett.*, 55(3), pp. 184–192, 2006.
2. G. Herrmann, M. C. Turner, I. Postlethwaite, and G. Guo, Practical implementation of a novel anti-windup scheme in a HDD-dual-stage servo-system, *IEEE/ASME Trans. Mech.*, 9(3), pp. 580–592, 2004.
3. C. Du, T. Gao, C. P. Tan, J. Yang, and L. Xie, Saturation control for an MTA based dual-stage actuation system, *IEEE Trans. Magn.*, 49(6), pp. 2526–2529, 2013.
4. G. Guo, D. Wu, and T. C. Chong, Modified dual-stage controller for dealing with secondary-stage actuator saturation, *IEEE Trans. Magn.*, 49(6), pp. 3587–3592, 2003.

5. X. Huang, R. Horowitz, and Y. Li, A comparative study of MEMS microactuator for use in a dual-stage servo with an instrumented suspension, *IEEE/ASME Trans. Mech.*, 11(5), pp. 524–532, 2006.
6. M. T. White, P. Hingwe, and T. Hirano, Comparison of a MEMS microactuator and a PZT milliactuator for high-bandwidth HDD servo, *Proc. of the 2004 American Control Conference*, Boston, MA, June 30–July 2, 2004, pp. 541–546.

7

Time Delay and Sampling Rate Effect on Control Performance of Dual-Stage Actuation Systems

7.1 Introduction

In a real-time digital control system, it takes time to implement data acquisition from sensors and signal processing to calculate and generate control commands. Both of them lead to time delay in the feedback control system and are, respectively, called signal measurement time delay and controller computing time delay. To implement the digital control system, sampling rate must be selected carefully and should be determined by not only satisfying the conditions of Shannon's sampling theorem [1] but also making the desired performance achievable. A higher sampling rate is generally preferred since it implies improved performance and better vibration rejection capability. However, increasing the sampling rate will make the time delay affect the control system performance more significantly. The time delay is usually smaller than the sampling interval in the digital control system. A longer time delay will have more severe impact on the control system performance.

If the signal measurement delay and the controller computing delay are hardly varied and can be calculated prior to putting the control system in use, they can be easily considered in the controller design. Since the delay is known beforehand and can be accommodated in the controllers, generally it will not cause stability problem for the control system. But the control performance due to the delay will be affected and thus needs to be paid much attention, particularly for those control systems which are expected to reach the stringent nanometer level positioning accuracy [2,3]. As the dual-stage actuation systems have been extensively used to push the control bandwidth, its control performance is severely affected by time delay and sampling rate [4,5], and in this chapter we particularly investigate the time delay and the sampling rate effect on the dual-stage actuation system.

89

7.2 Modeling of Time Delay

Theoretically, delay cannot be expressed as a rational polynomial. But there are some approximations for delay to be expressed as rational polynomial forms, among which Pade approximation is most well known for the controller design. The general form of Pade approximation is given by [6]

$$e^{-s\tau} = \frac{\text{num}(s)}{\text{den}(s)} \tag{7.1}$$

where

$$\text{num}(s) = \sum_{k=0}^{p} \frac{(p+q-k)!p!}{(p+q)!k!(p-k)!}(-s\tau)^k \tag{7.2}$$

and

$$\text{den}(s) = \sum_{k=0}^{q} \frac{(p+q-k)!q!}{(p+q)!k!(q-k)!}(s\tau)^k. \tag{7.3}$$

Equation (7.1) can be made as accurate as needed by choosing the orders p and q that are large enough. In general, (7.1) is an all-pass filter so that it will not affect the gain of the plant, and thus $p = q$.

Typically, the first-order Pade approximation is known as follows:

$$e^{-s\tau} \cong \frac{1-0.5\tau s}{1+0.5\tau s}. \tag{7.4}$$

To be as valid as possible for a particular design, the approximation should accurately follow the Bode plot of $e^{-s\tau}$ over the range of frequencies corresponding to the bandwidth of the designed system. We take $10\,\mu s$ delay as an example. Its first-order and tenth-order approximations are compared in Figure 7.1. For a control system which targets up to $10\,kHz$ bandwidth, the first-order approximation is accurate enough. Therefore, due to its simple form and accurate approximation, the first-order Pade approximation (7.4) is used to model the delay for the controller design.

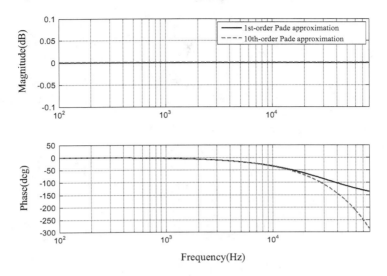

FIGURE 7.1
Pade approximation Bode plot comparison between the first and tenth orders for 10 µs delay. (From Ref. [2].)

7.3 Dual-Stage Actuation System Modeling with Time Delay for Controller Design

The dual-stage servo loop in hard disk drives is taken as an example for the investigation of time delay and sampling rate effect on control system performance [2]. Due to the processing of embedded burst signal for position error signals (PESs), the servo control loop is subjected to error signal measurement time delay including the burst signal duration, ADC (Analog to Digital Converter) conversion time, PES calculation time, etc., in addition to the controller computing time delay, as shown in Figure 7.2, where τ_m and τ_c are, respectively, used to denote the error signal measurement delay and

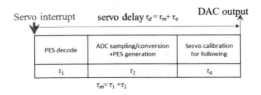

FIGURE 7.2
Servo delay in the servo control systems. (From Ref. [2].)

the controller computing delay with the effect of ZOH (zero-order-hold) included. The delays are included in the dual-stage servo control loop as shown in Figure 7.3, where $\bar{u}_p$ is the saturation level of the control input of the secondary actuator.

Figures 7.4 and 7.5, respectively, show the measured and the modeled frequency responses of the VCM (voice coil motor) actuator $P_v(s)$ and the PZT (Pb-Zr-Ti) milliactuator $P_p(s)$, which are combined to form the dual-stage actuation system and used to evaluate the effect of the time delay and the sampling rate on the control system performance.

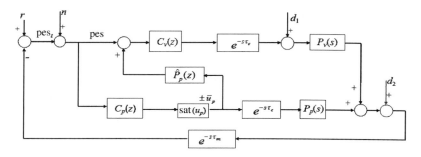

FIGURE 7.3
Block diagram of dual-stage servo control loop with PES measurement delay τ_m and controller computing delay τ_c. (From Ref. [2].)

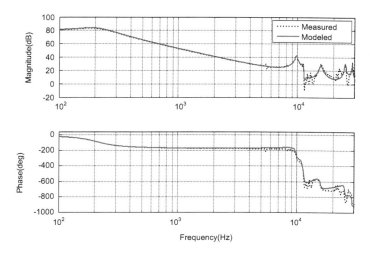

FIGURE 7.4
Frequency responses of the VCM actuator $P_v(s)$. (From Ref. [2].)

Effect of Time Delay and Sampling Rate

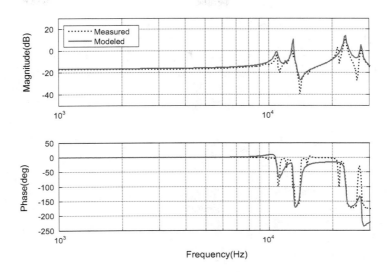

FIGURE 7.5
Frequency responses of the PZT milliactuator $P_p(s)$. (From Ref. [2].)

7.4 Controller Design with Time Delay for the Dual-Stage Actuation Systems

Denote $H_m(z)$ and $H_c(z)$ as the discretized time delay model for the delay τ_m and τ_c, respectively, and

$$H(z) = H_m(z)H_c(z) \tag{7.5}$$

From Figure 7.3, the open-loop transfer function of the whole system is derived as

$$G(z) = G_v(z) + G_p(z) + G_v(z)G_p(z) \tag{7.6}$$

where $G_v(z) = H(z)P_v(z)C_v(z)$ is the open-loop transfer function of the VCM actuator loop, and $G_p(z) = H(z)P_p(z)C_p(z)$ is the open-loop transfer function of the PZT milliactuator loop. As discussed in Chapter 2, the overall sensitivity function is given by

$$S(z) = \frac{1}{1+G(z)} = S_v(z)S_p(z) \tag{7.7}$$

where $S_v(z) = \dfrac{1}{1+G_v(z)}$ and $S_p(z) = \dfrac{1}{1+G_p(z)}$, which means $S(z)$ is the product of the sensitivity functions of the two individual loops. As such, the VCM and PZT controllers can be designed separately, which simplifies the overall control system design.

The used method to design the controllers $C_v(z)$ and $C_p(z)$ is the H_∞ loop shaping method stated in Chapter 2, which is applied to the plants $H(z)P_v(z)$ and $H(z)P_p(z)$, respectively. Figure 7.6 shows the frequency responses of $G_v(z)$ and $G_p(z)$. And, Figure 7.7 shows $|S_v(z)|$, $|S_p(z)|$, and $|S(z)|$ in the frequency domain.

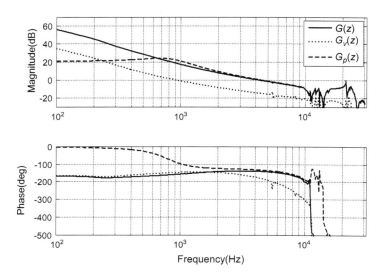

FIGURE 7.6
Frequency responses of the open loops $G_v(z)$, $G_p(z)$, and $G(z)$. (From Ref. [2].)

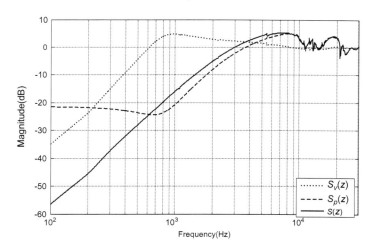

FIGURE 7.7
Magnitudes of sensitivity functions $S_v(z)$, $S_p(z)$, and $S(z)$. (From Ref. [2].)

Effect of Time Delay and Sampling Rate

The servo performances such as gain margin/phase margin, servo bandwidth, and the sensitivity function are mainly discussed in this chapter, since the gain/phase margins reflect the system stability, and the servo bandwidth and the sensitivity function imply the vibration rejection capability of the control system, which are the main concerns for the positioning accuracy.

7.5 Time Delay Effect on Dual-Stage System Control Performance

In this section, we assume the sampling interval $T_s = 16\,\mu s$ or the sampling rate is 62.5 kHz. We consider the effect of the PES measurement delay τ_m and the controller computing delay τ_c on control performance.

7.5.1 $(\tau_m + \tau_c) < T_s$

The time delay is required to be less than the sampling interval in most of the control systems, i.e., $(\tau_m + \tau_c) < T_s$, so that the control performance is not degraded too much by the time delay.

The controller computing delay τ_c is assumed to be 3 μs. Three values for τ_m are selected for investigation, i.e., $\tau_m = 3, 8,$ and 12 μs. The total delay time $(\tau_m + \tau_c) < 16\,\mu s = T_s$.

The control performances are compared in Table 7.1 among the three cases. The bandwidth is obviously reduced, especially when $\tau_m = 12\,\mu s$. Both stability margins tend to become less. Although a little change is seen in the gain margin for $\tau_m = 8$ and 12 μs, the bandwidth and the phase margin change apparently, which is also observed in Figure 7.8. These lead to the sensitivity function change, as seen in Figure 7.9, where the error rejection capability below 5 kHz is negatively affected as the delay is longer.

Consider the case of $\tau_m = \tau_c = 3\,\mu s$ with the results listed in Table 7.1. The total delay is reduced by 1 μs, respectively, in the measurement delay τ_m and the controller computing delay τ_c. The results are compared in Table 7.2. It is first mentioned that, as expected, 5 μs delay gives better performance than

TABLE 7.1

Servo Control Performance for Different Sensor Signal Measurement Delays and Controller Computing Delay $\tau_c = 3\,\mu s$

Sensor signal measurement delay τ_m (μs)	3	8	12
Servo bandwidth (kHz)	4.21	4.16	3.90
Gain margin (dB)	7.8	6.1	6.0
Phase margin (deg)	41.1	40.5	39.0

Source: From Ref. [2].

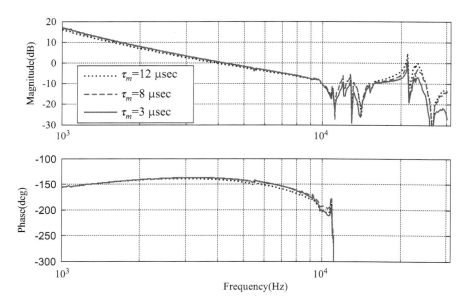

FIGURE 7.8
Open loop comparison for different sensor signal measurement delay τ_m and controller computing delay $\tau_c = 3\,\mu s$ (0 dB crossover frequency, i.e., servo bandwidth, is obviously reduced when $\tau_m = 12\,\mu s$). (From Ref. [2].)

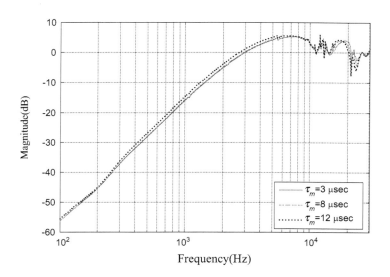

FIGURE 7.9
Sensitivity function comparison for different sensor signal measurement delay τ_m and controller computing delay $\tau_c = 3\,\mu s$ (($\tau_m + \tau_c) < T_s$). (From Ref. [2].)

TABLE 7.2
Servo Control Performance Comparison for Different Delay Combinations

(τ_m, τ_c) (μs)	(3,3)	(3,2)	(2,3)
Servo bandwidth (kHz)	4.21	4.33	4.36
Gain margin (dB)	7.8	7.3	6.8
Phase margin (deg)	41.1	39.8	43.7

Source: From Ref. [2].

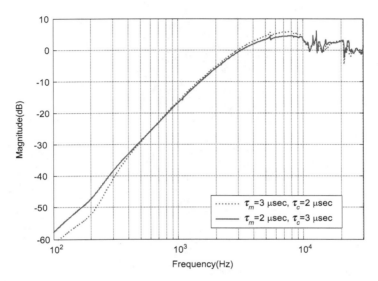

FIGURE 7.10
Sensitivity function comparison between (τ_m, τ_c) = (3,2) μs and (2,3) μs. (From Ref. [2].)

6 μs delay. Second, a more interesting finding is that reducing the measurement delay is more significant in terms of performance improvement than reducing the controller computing delay, as seen in Table 7.2 and Figure 7.10 where (τ_m, τ_c) = (2, 3) leads to a noticeable higher phase margin and therefore a lower hump of the sensitivity function.

7.5.2 $T_s < (\tau_m + \tau_c) < 2T_s$

In some particular cases, due to the limitation from developed systems, the delay has to be larger than the sampling interval. Table 7.3 lists three cases where the total delay is larger than T_s but less than $2T_s$. Although the system can be stabilized due to the prior known time delay involved in the controller design, the performance is much degraded, compared to Table 7.1. One obvious degradation is the bandwidth reduction, and another one is the much-reduced phase margin. It is also seen in Table 7.3 that both of them are further degraded as the delay time becomes longer. Figures 7.11 and 7.12

TABLE 7.3

Servo Control Performance for Different Sensor Signal Measurement Delays and Controller Computing Delay $\tau_c = 3\,\mu s$

Sensor signal measurement delay τ_m (µs)	13	18	22
Servo bandwidth (kHz)	3.85	3.64	3.44
Gain margin (dB)	6.02	5.92	6.36
Phase margin (deg)	38.8	36.8	35.9

Source: From Ref. [2].

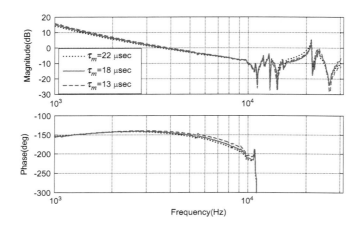

FIGURE 7.11
Open loop comparison for different PES measurement delay τ_m and controller computing delay $\tau_c = 3\,\mu s$ $((\tau_m + \tau_c) > T_s)$. (From Ref. [2].)

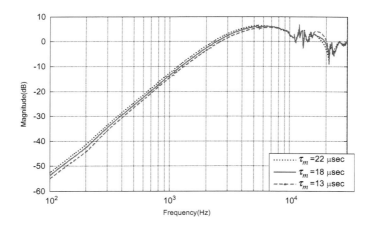

FIGURE 7.12
Sensitivity function comparison for different PES measurement delay τ_m and controller computing delay $\tau_c = 3\,\mu s$ $((\tau_m + \tau_c) > T_s)$. (From Ref. [2].)

Effect of Time Delay and Sampling Rate

show the progressive change of the open loop and the sensitivity function with the delay time. Due to the reduced phase margin, the hump of the sensitivity function in Figure 7.12 becomes higher as the delay is longer.

7.6 Sampling Rate Effect on Dual-Stage System Control Performance

In the previous section, the sampling rate is 62.5 kHz. In this section, we consider a lowered sampling rate, i.e., 40 kHz. The time delays are $\tau_m = 3\,\mu s$ and $\tau_c = 3\,\mu s$. As seen in Table 7.4, the lowered sampling rate causes much more degraded performance than the longer time delay (Table 7.1) and more even than the time delay over the sampling interval (Table 7.3). Figures 7.13 and 7.14 show the open-loop frequency responses and the sensitivity functions, respectively, designed with the two sampling rates. As observed in Figure 7.14, the sensitivity function designed with a higher sampling rate has vibration rejection capability in a wider frequency range and a lower hump. Another advantage due to high sampling rate in terms of control loop design is that it gives us more freedom to shape the sensitivity function of the control loop.

The above analysis implies that if the sampling rate can be increased markedly, it will have remarkable benefit for the servo control performance, given by tolerable time delay. And particularly for hard disk drives, the time delay can be estimated as a priori knowledge and has little variation, and thus can be considered in controller design so that it can be compensated for by the controller. However, if the time delay is too large, the system will not be able to achieve higher bandwidth even if the sampling rate is increased.

For the servo control in hard disk drives, the sampling rate is determined by the spindle motor speed in RPM (rotations/min) and the servo sector number and equals to

$$\frac{RPM}{60} \times Number\ of\ servo\ sector \tag{7.8}$$

TABLE 7.4

Servo Control Performance with Different Sampling Rates, Sensor Signal Measurement Delay $\tau_m = 3\,\mu s$, and Controller Computing Delay $\tau_c = 3\,\mu s$

Sampling rate	62.5	40
Servo bandwidth (kHz)	4.21	3.12
Gain margin (dB)	7.75	5.78
Phase margin (deg)	41.1	35.1

Source: From Ref. [2].

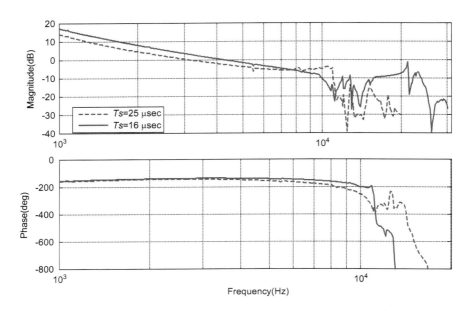

FIGURE 7.13
Open-loop frequency responses comparison with sampling rates 40 kHz ($T_s = 25\,\mu s$) and 62.5 kHz ($T_s = 16\,\mu s$). (From Ref. [2].)

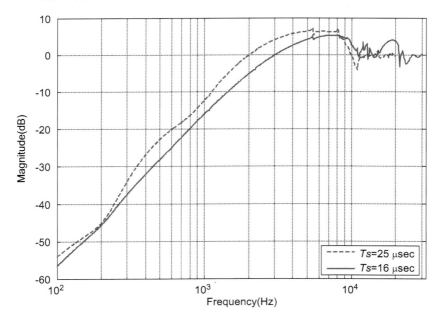

FIGURE 7.14
Sensitivity function comparison between the two cases of sampling rates 40 kHz ($T_s = 25\,\mu s$) and 62.5 kHz ($T_s = 16\,\mu s$). (From Ref. [2].)

according to which a higher sampling rate will need either a faster rotational speed or more servo space. However, faster rotational speed will cause higher vibration leading to worse servo positioning accuracy, and more servo space will sacrifice data space. The dedicated servo recording system, which is detailed in [7,8], provides a high sampling rate and has the advantage that it does not need to sacrifice disk surface for data storage. One practically applicable sampling rate is as high as 81 kHz. To match the servo recording system having such a high sampling rate, the PZT microactuator with the frequency responses in Figure 5.1 has been used. With the microactuator and the high sampling rate, the servo control loop can be designed to achieve up to 10 kHz control bandwidth.

Figure 7.15 shows the frequency responses of $G_v(z)$, $G_p(z)$, and $G(z)$, where it is seen that the servo bandwidth of 10 kHz is achieved. The sensitivity functions are plotted in Figure 7.16. It is seen that the 10 kHz dual-stage servo system has more than 10 dB attenuation ability to the vibrations within 4 kHz, and it is able to reject the vibrations in the frequency range up to 8 kHz.

If the vibration concentrates in 1–5 kHz, as shown in Figure 7.17, such as the audio vibration from speakers [9], the 10 kHz bandwidth servo loop makes a big difference and significantly outperforms the 5 kHz servo loop. This can be seen in Figure 7.18, where the error signal 3σ value versus the vibration magnitude is plotted for 5 kHz and 10 kHz dual-stage servo loops and a single-stage servo loop. When the error signal 3σ value is 15% with 5 kHz

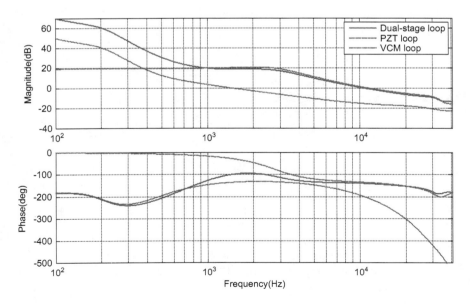

FIGURE 7.15
Open-loop frequency responses (VCM loop: $G_v(z)$; PZT loop: $G_p(z)$; and dual-stage loop: $G(z)$). (From Ref. [2].)

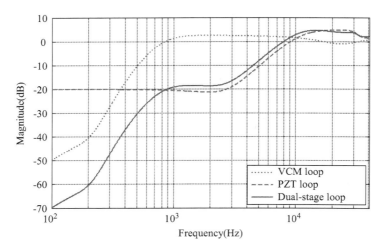

FIGURE 7.16
Sensitivity functions (VCM loop: $S_v(z)$; PZT loop: $S_p(z)$; and dual-stage loop $S(z)$. (From Ref. [2].)

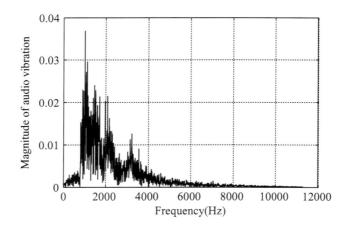

FIGURE 7.17
Spectrum of the audio vibration contained in PES. (From Ref. [2].)

servo loop, it is much better and below 6% with 10 kHz servo loop. However, for single-stage servo loop, it is tough to deal with such kind of vibrations. As observed in Figure 7.18, the error signal 3σ is already more than 15% tracks, even when the vibration is only one-fifth of the assumed full level. This is due to its low bandwidth and sensitivity function magnitude of more than 0 dB from 1 kHz upwards, leading to this higher-frequency vibration amplified by the servo loop.

Effect of Time Delay and Sampling Rate

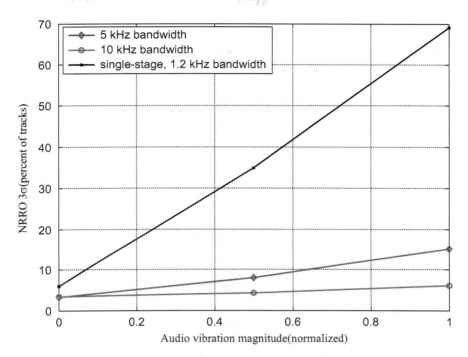

FIGURE 7.18
Dedicated servo error signal 3σ versus audio vibration (the level of the vibration causing 15% 3σ with 5 kHz servo loop is regarded as a full level and normalized as 1). (From Ref. [2].)

7.7 Conclusion

The sensor signal measurement delay and the controller computing delay have been modeled and involved in the controller design for the dual-stage actuation system. Two cases when the delay is within and over the sampling interval have been investigated. It has been shown that the control performance, such as stability margins and bandwidth, is degraded as the delay is longer. For the sensitivity function, as the delay is higher, the first 0 dB crossover frequency is lowered and the hump is higher.

Two sampling rates, 62.5 kHz and 40 kHz, have been used to study the sampling rate effect on the control performance. This increased sampling rate together with the reasonable delay time has produced significantly better control performance and vibration rejection capability. Moreover, with 81 kHz sampling rate, it has been shown that an ultra-high bandwidth of 10 kHz can be achieved for the dual-stage actuation system with a PZT microactuator.

References

1. M. Unser, Sampling-50 years after Shannon, *Proc. IEEE*, 88(4), pp. 569–587, 2000.
2. C. Du, A. Kong, and Y. Zhang, Time delay and sampling rate effect on dual-stage servo control performance, *Microsyst. Technol.*, 22(6), pp. 1213–1219, 2016.
3. R. Wood, M. Williams, A. Kavcic, J. Miles, The feasibility of magnetic recording at 10 Terabits per square inch on conventional media, *IEEE. Trans. Magn.*, 45(2), pp. 917–923, 2009.
4. M. T. White and W.-M. Lu, Hard disk drive bandwidth limitations due to sampling frequency and computational delay, *Proceedings of the 1999 IEEE/ASME International Conference on Advanced Intelligent Mechatronics*, September 19–23, 1999, Atlanta, US.
5. D. Abramovitch, T. Hurst, and D. Henze, The PES pareto method: uncovering the strata of position error signals in disk drives, In *Proc. of the 1997 American Control Conference*, Albuquerque, NM, June 3–6 1997, pp. 2888–2895.
6. G. H. Golub and C. F. Van Loan, *Matrix Computations*, Johns Hopkins University Press, Baltimore, 1989.
7. C. Du, Z. Yuan, A. Kong, and Y. Zhang, High-precision and fast response control for complex mechanical systems—servo performance of dedicated servo recording systems, *IEEE Trans. Magn.*, 53(3), p. 3101005, 2017.
8. Z. Yuan, J. Shi, C. L. Ong, P. S. Alexopoulos, C. Du, et al., Dedicated servo recording system and performance evaluation, *IEEE Trans. Magn.*, 51(4), p. 3100507, 2015.
9. J. Q. Mou, F. Lai, I. B. L. See, and W. Z. Lin, Analysis of structurally transmitted vibration of HDD in notebook computer, *IEEE Trans. Magn.*, 49(6), pp. 2818–2822, 2013.

8

PZT Hysteresis Modeling and Compensation

8.1 Introduction

Hysteresis nonlinearity is an inherent characteristic of PZT (Pb-Zr-Ti) materials. It decreases the performance of the control system and even results in limit cycle oscillations [1]. In [2], a general hysteresis definition is given by using rate-independent memory effects. For the hysteresis system, the output of the system depends not only on the present input but also on the past input or memory. Rate-independent means that the input-output relationship is invariant with respect to the change in time scales. The basic idea to deal with the hysteresis is through a compensation using the inverse of a right hysteresis model. So, the challenge is the PZT hysteresis modeling. In this chapter, the PZT milliactuator discussed in Chapter 1 is considered. The hysteresis model due to the PZT elements will be identified and used to design the compensator, and the dynamic responses of the structure with the hysteresis effects removed will be measured and identified to design the controller.

Based on the hysteresis model, a hysteresis compensator is designed and the effect of the hysteresis on the frequency response of the PZT-actuated structure is shown experimentally. A linear system can be identified by cascading the compensator with the PZT-actuated structure, and it can be seen that there is remarkable phase-lead in frequency responses by using the compensator, which is beneficial to the feedback controller design.

8.2 Modeling of Hysteresis

There are two models most widely used to describe the hysteresis behavior: phenomenological model such as Presaich [3] and Prandtl-Ishlinskii (PI) models [4], and physics-based models such as Duhem [2] and Bouc-Wen models [2]. Compared to the physics-based model, the phenomenological model allows a more precise modeling of the hysteresis behavior as it is created

105

based on the input/output data that can cover both the major and minor curves of the hysteresis behavior. PI model is a subset of Preisach model and its main advantage over Preisach model is that it is less computationally demanding to invert for feedforward control since an analytic inverse exists [5]. In what follows, the PI model and a generalized PI (GPI) model will be briefly introduced.

8.2.1 PI Model

The classical PI model is described as [6]

$$H(t) = qv(t) + \int_0^R p(r) F_r[v](t) dr \qquad (8.1)$$

where $v(t)$ is the input and $H(t)$ is the output, q is a positive constant, $p(r)$ is a density function satisfying $p(r) \geq 0$, and $F_r[v](t)$ is the hysteresis play operator shown in Figure 8.1 and described as follows.

For any piecewise monotone input function $v: [0, t_E] \to R$, $w(t) = F_r[v](t)$ is given by

$$\begin{aligned} w(0) &= f_r(v(0), 0) \\ w(t) &= f_r(v(0), w(t_i)) \text{ for } (t_i < t < t_{i+1}, 0 \leq i \leq N-1) \end{aligned} \qquad (8.2)$$

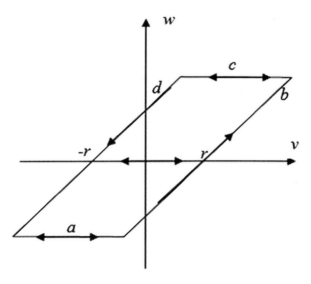

FIGURE 8.1
Hysteresis play operator.

PZT Hysteresis Modeling and Compensation

with

$$f_r(v,w) = \max(v-r, \min(v+r, w)),\tag{8.3}$$

where $0 = t_0 < t_1 < \ldots < t_N = t_E$ is a partition of $[0, t_E]$ such that the function v is monotone on each of the subintervals $[t_i, t_{i+1}]$.

According to (8.3) as well as Figure 8.1, when input v increases/decreases and $v - r \leq w \leq v + r$, output w of the play operator will assume the value of w and change along a or c branch depending upon its previous values. When v increases and $v - r > w$ and $w \leq v + r$, output w of the play operator will assume the value of $v - r$ and increase along the b branch. On the other hand, when v decreases and $v - r \leq v + r < w$, output w of the play operator will assume the value of $v + r$ and decrease along the d branch.

It can be seen from (8.1) that the PI model is an integral of many elementary play operators, which are parameterized by the threshold value r. A more complex hysteresis behavior can be modeled when more play operators with different threshold values are used.

8.2.2 GPI Model

Classical PI model cannot be used to model either asymmetric hysteresis loops or saturated hysteresis output. In [7], a GPI model is thus proposed so that asymmetric hysteresis loops or saturated hysteresis output can be modeled, based on the use of a generalized play operator. The GPI model is given by

$$H(t) = q\gamma(v(t)) + \int_0^R p(r)F_r^\gamma[v](t)dr\tag{8.4}$$

with

$$\gamma(v(t)) = \begin{cases} \gamma_u(v(t)) & \text{if } \dot{v}(t) \geq 0 \\ \gamma_d(v(t)) & \text{if } \dot{v}(t) < 0 \end{cases}.\tag{8.5}$$

The main difference in the GPI model is the addition of the envelope function γ, which is made up of γ_u and γ_d. An increase in input v will cause the output of the GPI model to depend upon the function γ_u, and a decrease in input v will cause the output to depend upon the function γ_d. In the discrete form, the GPI model will be

$$H(k) = q\gamma(v(k)) + \sum_{j=1}^n p_j(r_j)F_{r_j}^\gamma[v](k)dr\tag{8.6}$$

with

$$\gamma(v(k)) = \begin{cases} \gamma_u(v(k)) & \text{if } \dot{v}(k) \geq 0 \\ \gamma_d(v(k)) & \text{if } \dot{v}(k) < 0 \end{cases}, \tag{8.7}$$

where $v(k)$ is the discrete input with $k = [0, 1, 2,..., N]$ and N is the total number of discrete samples. In (8.6), n is the number of generalized play operator described as

$$w(0) = F_{r_j}^{\gamma}[v](0) = f_{r_j}^{\gamma}(v(0), 0)$$

$$w(k) = F_{r_j}^{\gamma}[v](k) = f_{r_j}^{\gamma}(v(k), F_{r_j}^{\gamma}[v](k-1)) \tag{8.8}$$

with

$$f_{r_j}^{\gamma}(v, w) = \max(\gamma_u(v) - r_j, \min(\gamma_d(v) + r_j, w)). \tag{8.9}$$

From (8.9), it can be seen that the output of the generalized play operator is dependent upon the functions γ_u and γ_d. The density function $p_j(r_j)$ and the envelope functions $\gamma_u(v)$ and $\gamma_d(v)$ are

$$p_j(r_j) = \lambda e^{-\delta r_j}, \text{ where } r_j = \rho j \text{ and } j = 1, 2, 3,..., n,$$

$$\gamma_u(v) = \alpha v - \beta, \tag{8.10}$$

$$\gamma_d(v) = \varepsilon v - \eta,$$

where λ, δ, ρ, α, β, ε, and η are constants.

8.2.3 Inverse GPI Model

For hysteresis compensation, the inverse of the hysteresis model will be used as the hysteresis compensator. An exact inverse of the GPI can be computed and the discrete form of the inverse GPI is expressed as [7]

$$H^{-1}(k) = \gamma^{-1}\left(q^{-1}v(k)\right) + \sum_{j=1}^{n} \hat{p}_j F_{\hat{r}_j}[v](k)) \tag{8.11}$$

with

$$\gamma^{-1}(v(k)) = \begin{cases} \gamma_u^{-1}(v(k)) & \text{if } \dot{v}(k) \geq 0 \\ \gamma_d^{-1}(v(k)) & \text{if } \dot{v}(k) < 0 \end{cases} \tag{8.12}$$

and n in (8.11) is the number of generalized play operators for obtaining the inverse GPI, which is described as

$$w(0) = F_{\hat{r}_j}[v](0) = f_{\hat{r}_j}(v(0), 0)$$
$$w(k) = F_{\hat{r}_j}[v](k) = f_{\hat{r}_j}(v(k), F_{\hat{r}_j}[v](k-1))$$
(8.13)

with

$$f_{\hat{r}_j}(v, w) = \max(v - \hat{r}_j, \min(v + \hat{r}_j, w)).$$
(8.14)

The parameters are obtained as

$$\hat{r}_j = qr_j + \sum_{i=1}^{j-1} p_i(r_j - r_i)$$

$$\hat{p}_j = -\frac{p_j}{(q + \sum_{i=1}^{j} p_i)(q + \sum_{i=1}^{j-1} p_i)}.$$
(8.15)

8.3 Application of GPI Model to a PZT-Actuated Structure

In this section, the GPI model is used to model the hysteresis behavior of a PZT-actuated structure, which works as a milliactuator with a similar view in Figure 1.3. A hysteresis compensator will be designed to compensate for the hysteresis behavior using the inverse of the GPI model. With the hysteresis compensator, a feedback control will be further designed for the PZT-actuated structure to investigate the compensation effect.

The proposed model illustrated in Figure 8.2 is equivalent to the physical system of the PZT-actuated structure, and the identification method is given as follows.

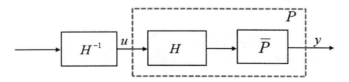

FIGURE 8.2
Hysteresis compensation scheme.

Step 1: A quasi-static excitation signal u is applied to the PZT suspension and then the output of the PZT-actuated structure y is measured.

Step 2: From the measured data (u, y), the subsystem H is modeled using GPI hysteresis model. The hysteresis compensator H^{-1} is designed based on the inverse of the subsystem H using (8.11)–(8.15).

Step 3: By cascading the inverse compensator H^{-1} with the PZT-actuated structure shown in Figure 8.2, the frequency response of the subsystem $\bar{P}$ can be measured and identified.

8.3.1 Modeling of the Hysteresis in the PZT-Actuated Structure

To identify the hysteresis behavior of the PZT-actuated structure, regular sinusoidal waves of different frequencies and amplitude 6 V are used as input signals into the PZT-actuated structure. The input and output signals (u, y) are, respectively, given to and collected from the PZT-actuated structure using a dSPACE 1104 system with a sampling frequency of 40 kHz. The hysteresis behavior of the PZT actuator at different frequencies is shown in Figure 8.3. Figure 8.4 shows the input signal that consists of regular sinusoidal waves of different frequencies and the corresponding output signal obtained from the PZT actuator. The parameters of the hysteresis model

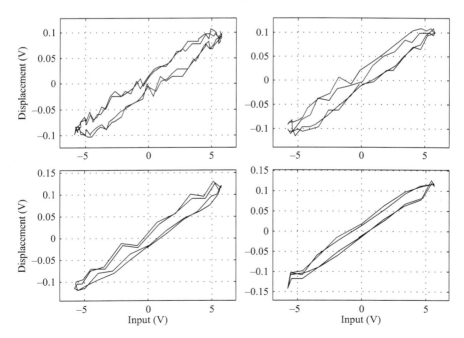

FIGURE 8.3
Measured hysteresis behavior for different frequencies. Top-left: 500 Hz. Top-right: 1 kHz. Bottom-left: 1.6 kHz. Bottom-right: 2 kHz.

PZT Hysteresis Modeling and Compensation

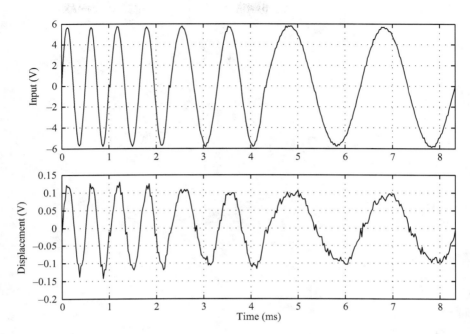

FIGURE 8.4
Input and corresponding output displacement signal used for hysteresis modeling.

described by (8.6)–(8.10) are obtained by doing curve fitting for the input-output data using nonlinear least-square optimization method. The number of generalized play operators, n, is chosen to be eight and the identified values of the GPI model parameters are shown in Table 8.1.

The comparison is shown in Figure 8.5 between the measured hysteresis and the hysteresis generated from the identified GPI model. Although the hysteresis behavior of the PZT actuator is measured at different frequencies,

TABLE 8.1

Parameters of GPI Model Identified Using Nonlinear Least-Squares Optimization

Parameters	Identified Values
α	0.0765
β	−0.0243
ρ	0.2889
λ	0.1116
δ	0.7086
q	0.2181
ε	0.0764
η	0.0248

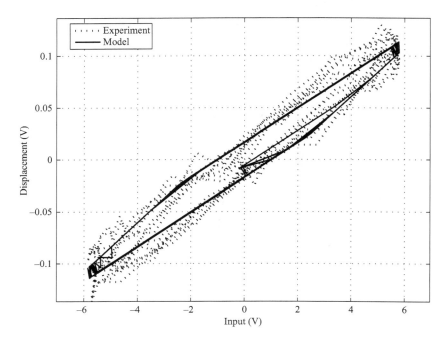

FIGURE 8.5
Comparison of measured hysteresis and hysteresis generated from GPI model.

the nonlinear least-square optimization is able to converge to a solution. It shows that the hysteresis has a rate independent characteristic, as it does not differ significantly at different frequencies.

8.3.2 Hysteresis Compensator Design

The hysteresis compensator is designed as the inverse GPI model in (8.11)–(8.15). The measured output signal shown in Figure 8.3 is fed into the open-loop scheme shown in Figure 8.2 and the inverse of the hysteresis behavior, which is generated using the inverse GPI model, is shown in the left of Figure 8.6. In the simulation, the hysteresis behavior of the PZT actuator is represented using the GPI model. The right of Figure 8.6 shows the hysteresis generated using the GPI model. Figure 8.7 showing the linear relationship of the input and the output of the system consisting of the hysteresis compensator and the GPI model means that the hysteresis can be completely compensated with the hysteresis compensator.

8.3.3 Experimental Verification

The effectiveness of both the open-loop hysteresis compensation structure shown in Figure 8.2 and the closed-loop structure shown in Figure 8.8 will

PZT Hysteresis Modeling and Compensation

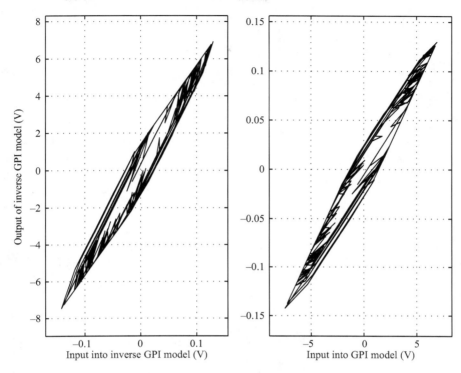

FIGURE 8.6
Inverse hysteresis behavior generated using inverse GPI model and hysteresis behavior generated using GPI model.

be verified by the experiment using dSPACE 1104 system with a sampling frequency of 40 kHz.

The open-loop hysteresis compensation structure is given input signals that are regular sinusoidal signals of different frequencies with amplitude of 0.1 V and the plots of the output signal against the input signal for the open-loop hysteresis compensation structure are shown in Figure 8.9. Using the same reference signals, simulation was carried out and the plots of the simulated output signal against the input signal for the open-loop hysteresis compensation structure are also included in Figure 8.9 for comparison. It can be observed that with the hysteresis compensator the hysteresis at different frequencies has been reduced significantly and the system consisting of the hysteresis compensator and the PZT actuator becomes approximately linear. Figure 8.10 shows the change of the actuator frequency responses due to the hysteresis compensation [8], which is an obvious phase-lead and results in faster step response, as verified in [9].

To further verify the hysteresis compensation effectiveness, a feedback controller is designed, which consists of a lag compensator combined with notch and peak filters discretized with a sampling frequency of 40 kHz. The

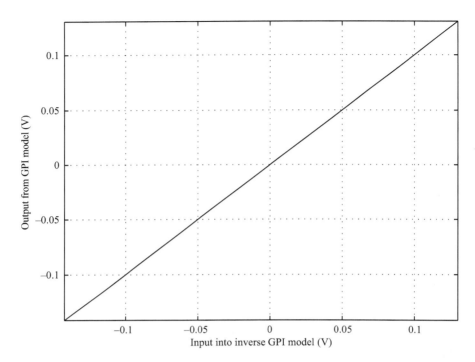

FIGURE 8.7
Simulated output against input signal for open-loop hysteresis compensation structure.

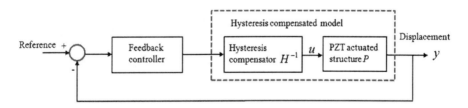

FIGURE 8.8
Feedback control structure with hysteresis compensation.

disturbance models in [10] are used. The closed-loop system with the disturbance models included is implemented using the equipment dSPACE 1104 system with a sampling frequency of 40 kHz and the error signal is measured using the equipment LDV (laser Doppler vibrometer).

Fast Fourier Transform (FFT) of the measured error signals was carried out and the spectra are shown in Figure 8.11. The cumulative sum of each spectrum, which is used to obtain 3σ value, is also shown in Figure 8.11. It is observed that with the use of the hysteresis compensation, the magnitude of the error spectrum at low frequencies is lower than that of the spectrum of

PZT Hysteresis Modeling and Compensation

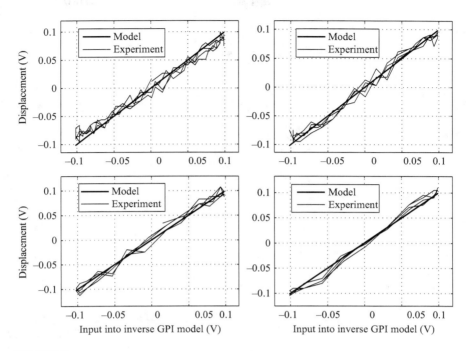

FIGURE 8.9
Measured and simulated output against input signal for open-loop hysteresis compensation structure. Top-left: 500 Hz. Top-right: 1 kHz. Bottom-left: 1.6 kHz. Bottom-right: 2 kHz.

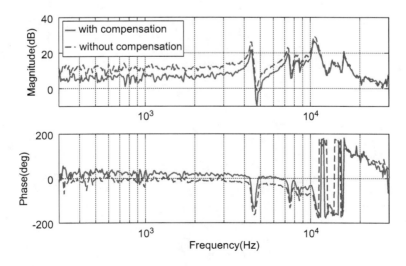

FIGURE 8.10
Comparison of PZT actuator frequency responses between with and without hysteresis compensation. (From Ref. [8].)

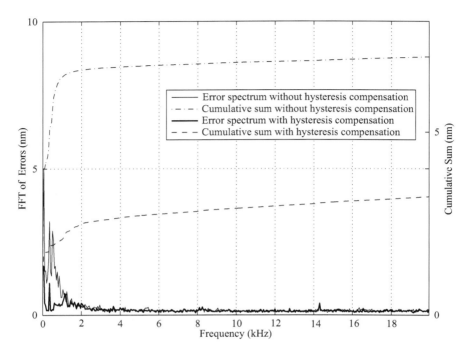

FIGURE 8.11
Comparison of FFT of errors for closed-loop system with and without hysteresis compensation.

the closed-loop system without the hysteresis compensation. In addition, the calculated 3σ value of the measured error with the hysteresis compensation is also much smaller than that without the hysteresis compensator [9]. It means that with the hysteresis compensation, the tracking performance or positioning accuracy of the PZT-actuated structure is improved.

8.4 Conclusion

The GPI model has been used to model the hysteresis behavior of the PZT-actuated structure from a dual-stage actuation system and its inverse to compensate for the hysteresis. The proposed control scheme has been able to compensate for the hysteresis behavior and results in the closed-loop system having a fast step response. In addition, the positioning accuracy of the PZT actuator has been improved. Although the GPI model has been applied to compensate for the hysteresis behavior in the PZT-actuated structure considered in this chapter, this method of hysteresis compensation can in fact be applied to other areas of science where hysteresis problem is encountered,

PZT Hysteresis Modeling and Compensation

as the modeling and compensation method using the GPI model is a mathematical approach without consideration of any underlying physical laws.

References

1. I. D. Mayergoyz, *Mathematical Models of Hysteresis*, Springer-Verlag, New York, 1991.
2. S. Rosenbaum, M. Ruderman, T. Strohla, and T. Bertram, Use of Jiles-Atherton and Preisach hysteresis models for inverse feed-forward control, *IEEE Trans. Magn.*, 46(12), pp. 3984–3989, 2010.
3. A. Cavallo, C. Natale, S. Pirozzi, and C. Visone, Effects of hysteresis compensation in feedback control systems, *IEEE Trans. Magn.*, 39(3), pp. 1389–1392, 2003.
4. K. Haiya, S. Komada, and J. Hirai, Tension control for tendon mechanisms by compensation of nonlinear spring characteristic equation error, *Proceedings of 11th IEEE International Workshop on Advanced Motion Control*, pp. 42–47, March 2010.
5. K. Y. Tsai and J. Y. Yen, Servo system design of a high-resolution piezo-driven fine stage for step-and-repeat microlithography systems, *Proceedings of IEEE IECON'99*, Vol. 1, pp. 11–16, 1999.
6. M. Brokate and J. Sprekels, *Hysteresis and Phase Transitions*, Springer, New York, 1996.
7. M. A. Janaideh, S. Rakheja, J. Mao, and C. Y. Su, Inverse generalized asymmetric Prandtl-Ishlinskii model for compensation of hysteresis nonlinearities in smart actuators, *Proceedings of International Conference on Networking, Sensing and Control*, pp. 834–839, March 2009.
8. Z. Zhang, C. Du, T. Gao, and L. Xie, Hysteresis modeling and compensation of PZT milliactuator in hard disk drives, *Proceedings of 2014 13th International Conference on Control, Automation, Robotics, and Vision (ICARCV)*, Singapore, December 10–12, 2014, pp. 980–985.
9. Y. Z. Tan, C. K. Pang, F. Hong, S. Won, and T. H. Lee, Hysteresis compensation of piezoelectric actuators in dual-stage hard disk drives, *Proc. of 2011 8th Asian Control Conference (ASCC)*, Kaohsiung, Taiwan, May 15–18, 2011, pp. 1024–1029.
10. F. Hong, C. K. Pang, W. E. Wong, and T. H. Lee, A peak filtering method with improved transient response for narrow-band disturbance rejection in hard disk drives, *Preprints of the 5th IFAC Symposium on Mechatronic Systems, MECHATRONICS2010–050*, Cambridge, MA, USA, September 13–15, 2010, pp. 71–75.

9

Seeking Control of Dual-Stage Actuation Systems with Trajectory Optimization

9.1 Introduction

In the implementation of the dual-stage actuation system, the primary actuator works as a first stage to generate coarse and slow movement, while the secondary actuator provides fine and fast positioning. With this servomechanism, the dual-stage actuators can improve both track-following and track-seeking performances for hard disk drives and outperform the single-stage actuator [1]. In the previous chapters, the control for high-precision positioning has been mainly addressed that is applicable for the track-following with the dual-stage actuation systems. This chapter is dedicated to the track-seeking control.

Most of the seeking controls for the dual-stage actuation systems are developed on the basis of 2DOF (two degrees of freedom) structure [2–4]. Reference [2] proposes three design steps of the feedforward controller that enables high-speed one-track seeking and the control method for short-span track-seeking over PZT secondary actuator's stroke limit (i.e., movement range). In [3], the seeking control scheme for the dual-stage actuation system uses two minimum jerk trajectories [5] with different preset seeking times to get reference inputs for the VCM (voice coil motor) primary actuator and the PZT (Pb-Zr-Ti) secondary actuator. In [4], the minimum jerk trajectory is used to compensate for nonzero initial values of the VCM states at the start of the settling phase. In addition to these 2DOF controls, several nonlinear control methods [6,7] are explored for the seeking control of dual-stage actuation servo systems, as well as PTOS (proximate-time-optimal-servomechanism) method [8].

An important issue in the 2DOF control is the generation of the reference trajectory or current profile for the VCM primary actuator and the secondary actuator. This chapter focuses on the optimization of reference trajectory or current profiles for the dual-stage actuators in order to have a fast and smooth seeking and small residue vibration.

9.2 Current Profile of VCM Primary Actuator

9.2.1 PTOS Method

As a conventional method, PTOS gives the current profile written as [9]

$$u_{ptos} = u_{max} \text{sat}\left(\frac{k_2(f(y_e(k)) - v(k))}{u_{max}}\right) \quad (9.1)$$

where $y_e(k) = r(k) - y(k)$ denotes the position error, $v(k)$ denotes the velocity, u_{max} is the maximum allowable input, and the function $f(\bullet)$ is given by

$$f(y_e(k)) = \begin{cases} \frac{k_1}{k_2} y_e(k), & |y_e(k)| \le y_l \\ \text{sgn}(y_e(k))g(k), & |y_e(k)| > y_l \end{cases} \quad (9.2)$$

with

$$g(k) = \sqrt{2u_{max} a\alpha |y_e(k)|} - \frac{u_{max}}{k_2}. \quad (9.3)$$

Figure 9.1 shows an example of PTOS current profiles.

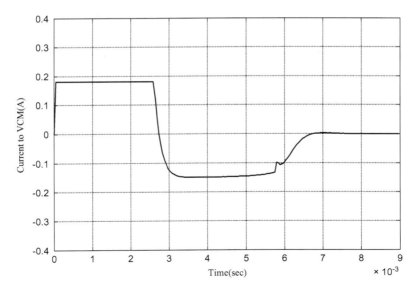

FIGURE 9.1
Current profile of VCM actuator from PTOS method.

Control of Dual-stage Actuation Systems

9.2.2 A General Form of VCM Current Profiles

The minimum-jerk current profiles of VCM actuators are known to be of the following polynomial form [5]:

$$a_v(k) = \text{scalar} \cdot \frac{r_t}{t_f^2}\left[60\frac{kT_s}{t_f} - 180\left(\frac{kT_s}{t_f}\right)^2 + 120\left(\frac{kT_s}{t_f}\right)^3\right], \quad k = 0,1,2,\ldots,\frac{t_f}{T_s} \quad (9.4)$$

where T_s and t_f are, respectively, the sampling rate and the desired settling time, and the scalar is determined according to the actuator plant gain. And, correspondingly, the reference is given by

$$r_v(k) = r_t\left[10\left(\frac{kT_s}{t_f}\right)^3 - 15\left(\frac{kT_s}{t_f}\right)^4 + 6\left(\frac{kT_s}{t_f}\right)^5\right] \quad (9.5)$$

where r_t is the seeking length.

Motivated by (9.5), we consider the following general form for a_v:

$$a_v(k) = c_0 + c_1\frac{kTs}{t_f} + c_2\left(\frac{kTs}{t_f}\right)^2 + c_3\left(\frac{kT_s}{t_f}\right)^3 + \cdots + c_p\left(\frac{kT_s}{t_f}\right)^p. \quad (9.6)$$

The current profile form in (9.6) gives much more freedom in terms of coefficients ($c_0, c_1, \ldots, c_p$) as variables to optimize certain performances. An example of the current profiles designed from the general form is shown in Figure 9.2. One of the applications is that this current profile is efficient in reducing the interaction between these types of two or multiple coupled actuation systems [10].

9.3 Control System Structure for the Dual-Stage Actuation System

Consider the dual-stage servo control system with a VCM actuator as the primary stage and a PZT milliactuator as the secondary stage. The models of the VCM actuator and the PZT milliactuator are partitioned as rigid part and flexible resonance part and, respectively, written as

$$P_v(s) = P_{vr}(s)P_{vf}(s) \quad (9.7)$$

with the rigid part $P_{vr}(s)$ and the resonance part $P_{vf}(s)$, and

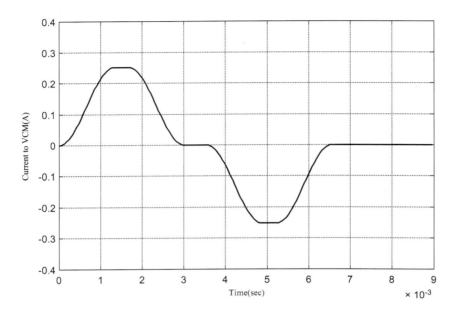

FIGURE 9.2
Current profile of VCM actuator from the high order Equation (9.6).

$$P_p(s) = K_p P_{pf}(s) \tag{9.8}$$

with the PZT gain K_p and the resonance part $P_{pf}(s)$.

The seeking control adopts the feedforward control structure as shown in Figure 9.3, where P_{vr} and K_p are the rigid parts of the plant models (9.7) and (9.8), a_v is the VCM current profile, and r_d is the reference trajectory of

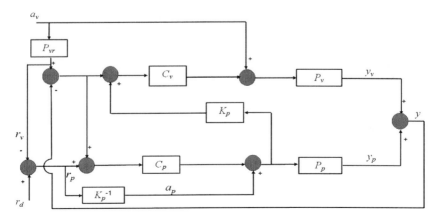

FIGURE 9.3
Block diagram of a seeking control structure for the dual-stage actuation system.

Control of Dual-stage Actuation Systems 123

the total dual-stage system. $C_v(z)$ and $C_p(z)$ are the track-following controllers designed with the method in Chapter 2. In the next section, we present the design of the VCM current profile a_v and the reference trajectory r_d, aiming to achieve a fast and smooth seeking response.

9.4 Design of VCM Current Profile a_v and Dual-Stage Reference Trajectory r_d

In Figure 9.3, the dual-stage system output y is derived as

$$
\begin{aligned}
y = &\left(1 - \frac{1}{1 + P_v C_v + P_p C_p + K_p P_v C_v C_p}\right) r_d \\
&+ \frac{P_v}{1 + P_v C_v + P_p C_p + K_p P_v C_v C_p} a_v \\
&+ \frac{P_{pf}}{1 + P_v C_v + P_p C_p + K_p P_v C_v C_p}(r_d - P_{vr} a_v),
\end{aligned}
\tag{9.9}
$$

where the first item is due to the closed loop for the track-following process, the second item is due to the VCM current profile a_v, and the third item is actually attributed to the PZT reference $r_p = r_d - P_{vr} a_v$. When there are no flexible components in VCM and PZT, i.e., $P_{vf}(s) = 1$ and $P_{pf}(s) = 1$, $y = r_d$.

It is known that the minimum-jerk reference and current profile of the single-stage VCM actuator are designed as in (9.5) and (9.4). Ideally, if the dual-stage reference trajectory r_d is designed as in (9.5) with a shorter settling time t_d than t_f, i.e.,

$$
r_d(k) = r_t \left[10\left(\frac{kT_s}{t_d}\right)^3 - 15\left(\frac{kT_s}{t_d}\right)^4 + 6\left(\frac{kT_s}{t_d}\right)^5 \right],
\tag{9.10}
$$

the PZT reference r_p and the current profile a_p could be determined as

$$
r_p(k) = r_d(k) - r_v(k)
\tag{9.11}
$$

$$
a_p(k) = K_p^{-1} r_p(k).
\tag{9.12}
$$

However, used in the dual-stage system, r_v in (9.5) and r_d in (9.10) or r_p in (9.11) cannot guarantee a satisfactory seeking performance of the entire dual-stage actuation system. Therefore, we consider the general form of a_v in (9.6) and the following general forms for r_d:

$$r_d(k) = d_0 + d_1 \frac{kT_s}{t_d} + d_2 \left(\frac{kT_s}{t_d} \right)^2 + d_3 \left(\frac{kT_s}{t_d} \right)^3 + \cdots + d_q \left(\frac{kT_s}{t_d} \right)^q. \tag{9.13}$$

And, a performance function is defined as

$$J = \sqrt{\sum_{k=0}^{N} (y(k) - y_d(k))^2} \tag{9.14}$$

where $y_d(k)$ is a desired seeking response, which can be determined by (9.10). As such, we can state the objective as: Given the feedforward control structure as in Figure 9.3, we design the coefficients of the polynomials in (9.6) and (9.13) by optimizing the performance function (9.14).

Matlab function "fminsearch" can be used to conduct this optimization. However, it is found that without any constraint, the optimization could lead to an unstable case for the individual VCM and PZT actuators, although their combined position $y_v + y_p$ approaches y_d. Therefore, some constraints to VCM or PZT movement must be imposed on the optimization. One simple choice is that the VCM current profile is fixed as (9.4) and the optimization is carried out to obtain the dual-stage reference r_d.

Denote

$$T_{r_d y} = 1 - \frac{1}{1 + P_v C_v + P_p C_p + K_p P_v C_v C_p}$$

$$T_{a_v y} = \frac{P_v}{1 + P_v C_v + P_p C_p + K_p P_v C_v C_p}$$

$$T_{r_p y} = \frac{P_{pf}}{1 + P_v C_v + P_p C_p + K_p P_v C_v C_p}$$

and

$$r_p = r_d - P_{vr} a_v.$$

Equation (9.9) is rewritten as

$$y = T_{r_d y} r_d + T_{a_v y} a_v + T_{r_p y} r_p. \tag{9.15}$$

Control of Dual-stage Actuation Systems

The procedure of the optimization is then described as follows.

Step 1. Given the system as in Figure 9.3, calculate the transfer functions $T_{r_d y}$, $T_{a_v y}$, and $T_{r_p y}$. Matlab function "dlinmod" can be used to obtain the transfer functions.

Step 2. The initial values of the coefficients in (9.13) can be determined according to (9.10).

Step 3. Calculate y based on (9.15) using Matlab function "lsim" within the fixed time period $[0, t_d]$.

Step 4. Use Matlab function "fminsearch" to minimize J in (9.14) and it will arrive at final coefficients for r_d in (9.13).

Note that the reference r_d is generated according to (9.13) in $t \in [0, t_d]$, and after t_d it is discontinuously fixed as one, as seen in Figure 9.4. The same issue is seen for a_v in (9.4), where after t_f, $a_v = 0$. These shortcomings are not good for the whole system performance. As seen in Figure 9.5, a big oscillation is observed after reaching the target. As such, we propose a two-stage optimization which makes use of a low-pass filter $F(z)$ to generate a reference trajectory in the later stage so that a smooth reference trajectory r_d can be obtained. The reference r_d is then given by

$$r_d = \begin{cases} (4.10), & \text{for } t \leq t_d \\ F(z) \cdot r_t, & \text{for } t > t_d \end{cases}, \quad (9.16)$$

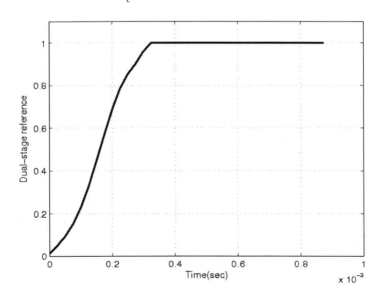

FIGURE 9.4
Reference trajectory r_d.

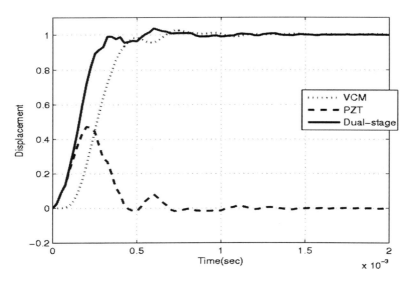

FIGURE 9.5
Displacements of VCM actuator (y_v) and PZT milliactuator (y_p), and the total $y = y_v + y_p$.

which is illustrated in Figure 9.6.
Denote

$$F(z) = \frac{b_m z^m + \cdots + b_1 z + b_0}{z^n + a_{n-1} z^{n-1} + \cdots + a_1 z + a_0} \tag{9.17}$$

which will be involved in the optimization to minimize the performance function (9.14). The two-stage optimization is thus summarized as follows.

1. Minimize J, in $0 \leq t \leq t_d$, and obtain the coefficients of r_d in (9.13).
2. Minimize J with $r_d = F(z) r_t$, in $t > t_d$, and obtain the parameters of $F(z)$ in (9.17).

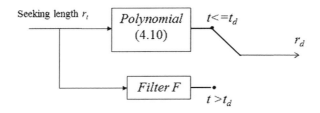

FIGURE 9.6
Generation of dual-stage reference r_d.

The second-stage optimization should be conducted in a sufficiently long period of time. This is to ensure the stability of the whole system as the system stability is much related to the second reference $F(z)r_t$.

9.5 Seeking within PZT Milliactuator Stroke

Assume that the stroke of the PZT milliactuator is 1 µm. The seeking length considered here is 1 µm. The optimization is completed after 15,000 iterations and has arrived at an eighth-order polynomial for r_d in (9.13). The filter $F(z)$ is obtained as

$$F(z) = \frac{0.01493z^2 + 0.02987z + 0.01493}{z^2 - 1.626z + 0.6855}. \tag{9.18}$$

The obtained VCM current profile and the reference trajectory r_d are shown in Figure 9.7, and the resultant seeking response is shown in Figure 9.8. It takes 0.3 ms for the positioning head to reach the target track with a small oscillation after reaching the target track. The low-pass filter $F(z)$ will cause some delay to the seeking response. This can be compensated partially by

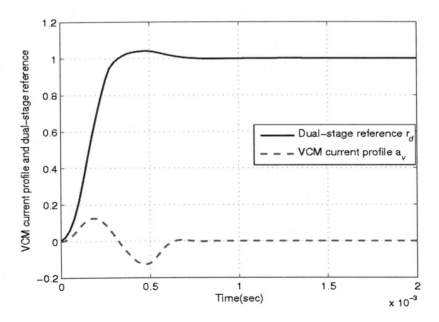

FIGURE 9.7
VCM current profile and dual-stage reference trajectory.

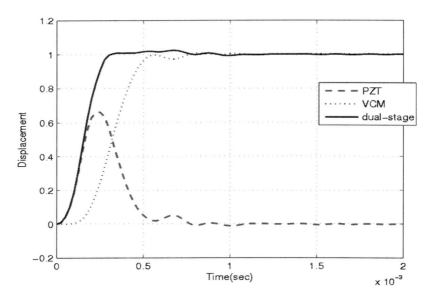

FIGURE 9.8
1 μm seeking response.

setting a smaller t_d for the polynomial generation of r_d than the target t_d to be reached ultimately.

Note that in Figure 9.3, $r_p = r_d - r_v$. One may prefer an optimization for r_p directly. But this will make it more difficult to obtain a satisfactory seeking response. The reason is that r_p, as in (9.11), is obtained from the polynomials r_d in (9.10) and r_v in (9.5) and is set abruptly to be zero after t_f, which makes it difficult to conduct the optimization in different time periods like that in (9.16). In view of this, the optimization in terms of r_p is not suggested using the optimization method proposed in the previous section.

9.6 Seeking over PZT Milliactuator Stroke

The problem for seeking over the PZT milliactuator stroke is how to generate a reference trajectory r_p within the PZT milliactuator stroke. In this section, we will make use of the previously obtained reference trajectories for seeking within PZT stroke to obtain the reference r_p for seeking over PZT stroke. In this case, we use the control structure shown in Figure 9.9.

Let the seeking length be denoted by l, and

$$r_m = l \cdot r_d - l \cdot P_{vr} a_v \tag{9.19}$$

Control of Dual-stage Actuation Systems

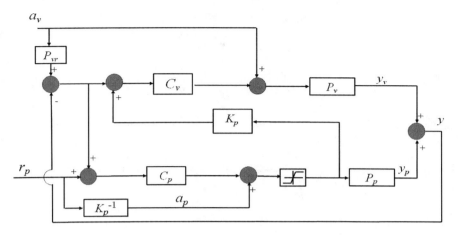

FIGURE 9.9
Seeking control structure with the seeking length more than PZT milliactuator stroke.

where r_d is obtained in the previous section for 1 μm seeking, and a_v takes the form as in (9.4). r_p can be designed as

$$r_p = \begin{cases} r_m, & \text{if } r_m \leq 1 \\ 1, & \text{if } r_m > 1 \end{cases}. \tag{9.20}$$

Using this reference r_p, if a big oscillation appears in the seeking response y, a notch filter can be used to reduce the oscillation. Then, the new r_p is that produced after passing through the notch filter. Moreover, we may need to tune t_f in a_v so that a suitable seeking time is acquired. With the method proposed here, to accomplish a seeking with $l = 3$ μm, the VCM current a_v and PZT reference r_p are displayed in Figure 9.10. The resultant seeking response is shown in Figure 9.11, which indicates that it takes 0.35 ms to reach the target with an oscillation within 3% of the seeking length.

9.7 Conclusion

This chapter has proposed an optimization method to generate the reference trajectory for the dual-stage actuation system to have a fast and smooth seeking and small residue vibration. The optimizations in different time periods have been carried out in the dual-stage servo loop with the predesigned track-following controllers. With the VCM current profile fixed as the minimum-jerk polynomial form, the beginning stage of the dual-stage

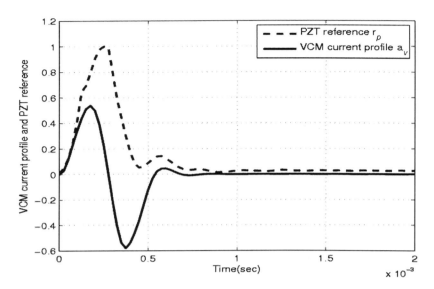

FIGURE 9.10
VCM current a_v and PZT reference r_p.

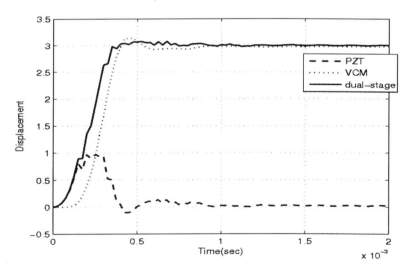

FIGURE 9.11
3 μm seeking response over PZT milliactuator stroke.

reference trajectory takes the form of a polynomial whose coefficients have been obtained through the optimization, and its later stage relies on the optimization for a low-pass filter. All steps of the optimizations to reach a desired dual-stage reference trajectory have been presented in detail. The application results in the seeking within the PZT secondary actuator's stroke

Control of Dual-stage Actuation Systems

and over its stroke have been presented and discussed to show the effectiveness of the optimization methods.

References

1. S. H. Lee, Y. H. Kim, and C. C. Chung, Dual-stage actuator disk drives for improved servo performance: track follow, track seek, and settle, *IEEE Trans. Magn.*, 37(4), pp. 1887–1890, 2001.
2. M. Kobayashi and R. Horowitz, "Track seek control for hard disk dual-stage servo systems", *IEEE Trans. Magn.*, 37(2), pp. 949–954, 2001.
3. J. Ding, F. Marcassa, and M. Tomizuka, Short seeking control with minimum jerk trajectories for dual actuator HDD systems, *Proc. of the 2004 American Control Conference*, Boston, Massachusetts, June 30–July 2, 2004, pp. 529–534.
4. H. Numasato and M. Tomizuka, Settling control and performance of a dual-actuator system for hard disk drives, *IEEE/ASME Trans. Mech.*, 8(4), pp. 431–438, 2003.
5. Y. Mizoshita, S. Hasegawa, and K. Takaishi, Vibration minimized access control for disk drives, *IEEE Trans. Magn.*, 32(3), pp. 1793–1798, 1996.
6. Y. Li, V. Venkataramanan, G. Guo, and Y. Wang, Dynamic nonlinear control for fast seek-settling performance in hard disk drives, *IEEE Trans. Ind. Electron.*, 54(2), pp. 951–962, 2007.
7. J. Zheng, M. Fu, Y. Wang, and C. Du, Nonlinear tracking control for a hard disk drive dual-stage actuator system, *IEEE/ASME Trans. Mech.*, 13(5), pp. 510–518, 2008.
8. B. Hredzak, G. Hermann, and G. Guo, A proximate-time-optimal-control design and its application to a hard disk drive dual-stage actuator system, *IEEE Trans. Magn.*, 42(6), pp. 1708–1715, 2006.
9. M. L. Workman, R. L. Kosut, and G. F. Franklin, Adaptive proximate time-optimal servomechanisms: Discrete-time case, *Proc. of the 26th Conference on Decision and Control*, Los Angeles, CA, US, December 1987, pp. 1548–1553.
10. J. C. Morris and Y. P. Hsin, Disturbance rejection of mechanical interaction for dual-actuator disc drives using adaptive feedforward servo, US patent No. 6493172B1, Seagate, 2002.

10

High-Frequency Vibration Control Using PZT Active Damping

10.1 Introduction

To achieve the goal of high precision positioning, the closed-loop control bandwidth has been increased accordingly for effectively suppressing the dominant vibrations through employing dual-stage or multi-stage actuation systems. However, the resonance modes of a flexible structure usually lie in a frequency range beyond the achievable control bandwidth, and these resonances are easily excited by exogenous sources and induce high-frequency structural vibrations. With the increased bandwidth and sometimes with further attenuation in the low-frequency range, these high-frequency vibrations tend to be amplified according to Bode's integral theorem [1–3]. As a consequence, these high-frequency vibrations will become a significant obstacle to achieving higher positioning accuracy.

There exist several techniques for dealing with the high-frequency structural vibrations. The commonly used one is to insert notch filters into the control loop to ensure the stability of control systems [4,5]. However, notch filters will reduce the phase margin and affect the system robustness. Besides, notch filters can only prevent the control input from exciting those vibration modes but cannot actively compensate for those vibrations. Furthermore, the achievable feedback control performance is limited by feedback control sampling rate and computational delay as discussed in Chapter 7, and thus these vibrations cannot be sufficiently attenuated with only feedback control. Therefore, the approach of using sensors placed on the flexible structure has been proposed. One idea using this sensor is to actively damp the resonances so as to reduce the effect of the structural vibrations on the positioning accuracy, which is implemented with a higher sampling rate than that of the feedback control loop.

As a typical application of high precision positioning control, the hard disk drives equipped with dual-stage actuation mechanism are needed to meet the high track density requirement. For instance, for an areal density of 10 Tb/in^2, the corresponding track density is about 1,600 k tracks/in, which implies a track width of 16 nm and an allowable 3σ track misregistration of 1.6 nm. To achieve this goal, the closed-loop servo bandwidth has to be increased accordingly for high positioning accuracy. However, with the extended servo bandwidth, airflow excited suspension structural vibrations [6,7], which usually lie in a frequency range beyond the control bandwidth, will be amplified and limit the positioning accuracy improvement. Instrumented suspensions have then been developed to help tackle this issue. In [8], an instrumented suspension with a strain gauge has been used to damp the resonance modes of a VCM (voice coil motor) actuator. In [9], butterfly resonance mode has been detected by a strain sensor and damped using the feedback signal from the sensor to feed into the VCM input. In [10], the PZT (Pb-Zr-Ti) component of the PZT actuated suspension has been used as a sensor for actively damping the VCM resonance modes. For the dual-stage actuation systems, actively damping resonance modes was proposed as early as in [11], which works as one way to push the servo bandwidth by damping the VCM resonance modes with an acceleration sensor, and in [12] using a PZT-actuated suspension doing self-sensing to damp VCM and PZT actuators' resonance modes. These approaches are able to attenuate the effect of the airflow induced structural vibrations.

In this chapter, first a singular perturbation method will be presented to design the damping controller using the PZT sensor signal. Later on, the design problem of the active damping controller is formulated as a H_2 control problem and a mixed H_2/H_∞ control problem, and the damping controller using both the methods is obtained by solving linear matrix inequalities (LMIs). Different from the LQG (linear quadratic Gaussian) method [8, 13], which acts as a special case of H_2 method, the H_2 damping control method is a general systematic method and does not need to tune cost function weighting matrices. For the mixed H_2/H_∞ method, in addition to the resonance modes detected by a sensor and to be damped, those not to be damped but detected by the sensor as well are taken into account and described as uncertainties. The proposed mixed H_2/H_∞ method is effective for active damping control and attractive from a practical point of view, considering the fact that a sensor picking up other unwanted modes is difficult to avoid due to the limitation of sensor location. Both of the proposed methods are applied to a VCM actuation mechanical system with a PZT sensor to detect the resonance modes for active vibration control.

10.2 Singular Perturbation Method-Based Controller Design

10.2.1 Singular Perturbation Control Topology

The singular perturbation design technique involves decomposing the system dynamics into slow and fast subsystems assuming that they operate on different time scales, which make them independent of each other [14,15] and allow independent controller design.

Consider the actuation structure with a VCM actuator mounted with PZT elements, shown in Figure 10.1. The singular perturbation control topology using the PZT as a sensor is shown in Figure 10.2. $\bar{P}_V$, $\tilde{P}_V$, and $\tilde{P}_V^*$ denote slow subsystems, fast subsystems, and estimates, respectively. The slow controller $\bar{C}_V$ and fast controllers ($\tilde{C}_V$ and $\tilde{P}_V^*$) operate at slow frequency $\bar{f}$ and fast frequency $\tilde{f} = \bar{f}/\varepsilon$ ($0 < \varepsilon \ll 1$), respectively. The fast subsystem dynamics $\tilde{y}_V$ are measured from the PZT structure as a sensor through a signal conditioning amplifier and estimated through $\tilde{P}_V^*$.

10.2.2 Identification of Fast Dynamics Using PZT as a Sensor

The PZT is employed solely as a sensor to detect high-frequency dynamics of the VCM actuated structure. The fast dynamics of $\tilde{P}_V^*$ should be estimated or measured for vibration rejection using inner loop compensation.

To identify the high-frequency dynamics and implement the control topology, a signal conditioning amplifier is used for the sensor signal. Using a similar methodology to that detailed in [16], swept sine excitation is injected

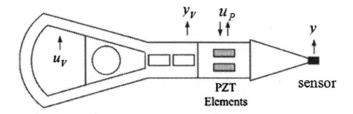

FIGURE 10.1
Illustrative diagram of a VCM actuator with mounted PZT elements showing input and output or measurement signals. (From Ref. [17].)

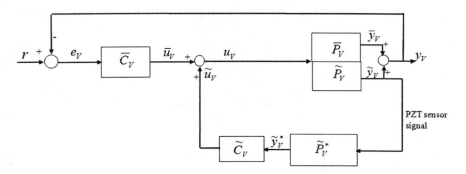

FIGURE 10.2
Block diagram of the singular perturbation-based control system.

at u_v, while y measured from the equipment LDV (laser Doppler vibrometer) is connected to Channel 1 and u_p from the PZT sensor through the amplifier is connected to Channel 2 of the equipment DSA (dynamic signal analyzer). The frequency response of the transfer function $T_{yu_P} = \dfrac{u_P}{y}$ from y to u_p is measured and shown in Figure 10.3, where it is seen that the estimation is effective at most frequencies and the frequencies of the antiresonant zeros correspond to the natural frequencies of the whole structure. The state estimator measured u_p from the amplifier to y can be identified with a noncausal transfer function. The high-frequency dynamics can hence be easily estimated online through a digital inverse of T_{yu_P} from the output of the amplifier.

10.2.3 Design of Controllers

According to the singular perturbation theory, the overall control signal u_v is given by $u_V = \bar{u}_V + \tilde{u}_V$, where $\bar{u}_V$ and $\tilde{u}_V$ are designed separately using the slow and the fast subsystems, respectively.

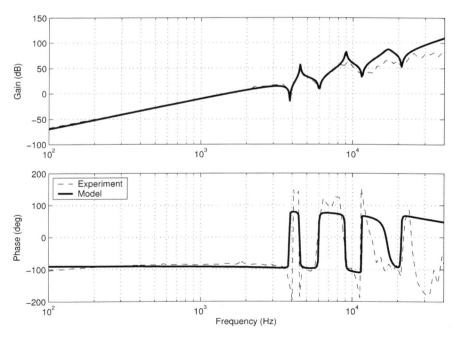

FIGURE 10.3
Frequency response of transfer function from y to u_p using PZT active suspension as a sensor. (From Ref. [17].)

High-Frequency Vibration Control

10.2.3.1 Fast Subsystem Estimator $\tilde{G}_V^*$

For fast inner loop compensation to be implementable, the high-frequency dynamics of the VCM and PZT active suspension should be available. By designing the fast subsystem estimator $\tilde{G}_V^*$ as the inverse of T_{yup}, the high-frequency dynamics estimate $\tilde{y}_V^*$ can be obtained online using the output of the amplifier (from the measured u_p) as input. This low-pass filter removes much measurement noise from the online estimator, which permits a low-order fast controller $\tilde{C}_V$ to be designed.

10.2.3.2 Fast Controller $\tilde{C}_V$

For the design in [17], a low-order lead compensator of the form

$$\tilde{C}_V(s) = \kappa \frac{s + 6283}{s + 1.257 \times 10^5} \tag{10.1}$$

is used as a fast controller $\tilde{C}_V$ with $1 < \kappa < 20$. $\tilde{C}_V$ is analogous to a high gain PD (proportional-derivative) control commonly used for controlling the robot manipulators [18]. A larger κ will increase the amount of active high-frequency vibration suppression, but the sensing noise will be accentuated accordingly and vice versa for a smaller value of κ. As such, a compromise should be considered when choosing κ so as to balance between the amount of suppression of high-frequency mechanical vibrations (and inner loop stability) and the amount of sensing noise. Here, κ is chosen as $\kappa = 10$.

10.2.3.3 Slow Controller $\bar{C}_V$

As the slow subsystem consists mainly of the rigid body modes (low-frequency double integrator), the controller $\bar{C}_V$ is designed as the lag-lead compensator recommended in [16] augmented with a low-pass filter and given by

$$\bar{C}_V = K_V \cdot \frac{s + \dfrac{2\pi f_V}{2\alpha}}{s + 2\pi 10} \cdot \frac{s + \dfrac{2\pi f_V}{\alpha}}{s + 2\alpha 2\pi f_V} \cdot \frac{\dfrac{\alpha\pi f_V}{2}}{s + \dfrac{\alpha\pi f_V}{2}} \tag{10.2}$$

where $5 < \alpha < 10$ is used. K_v is calculated by setting $\left|\bar{C}_V(j2\pi f_V)\bar{G}_V(j2\pi f_V)\right| = 1$, where $\bar{G}_V$ is the slow subsystem mainly determined by the rigid body dynamics, f_v is the gain crossover frequency of the compensated VCM open loop and is usually chosen at about one-fifth of the natural frequency of the first major resonant mode of the VCM for stability in the digital control systems. The first lag term in $\bar{C}_V$ is used to increase the low frequency gain for low frequency disturbance rejection and tracking accuracy in the slow outer loop.

The second term increases the phase margin of the outer open loop to ensure stability in the crossover region. An additional low-pass filter is cascaded in the third term of $\bar{C}_V$ to increase the high-frequency roll-off and stability of the stiffened VCM actuation structure due to a reduced relative degree after inner loop compensation.

10.2.4 Simulation and Experimental Results

10.2.4.1 Frequency Responses

The frequency response of the transfer function from u_v to y using PZT strip as a sensor with high-frequency inner loop compensation is shown in Figure 10.4. The flexible modes at 3.9, 6.0, 5.9, 11.5, and 21.0 kHz are effectively damped with a reduced relative degree using the proposed control scheme with the PZT used as an additional sensor. With this inner loop, it is expected to have a possible higher servo bandwidth and low sensitivity function. This is verified with the experimental sensitivity functions S of the proposed control scheme and the conventional notch filter-based control, shown in Figure 10.5, where $T = 1 - S$. The proposed control offers stronger

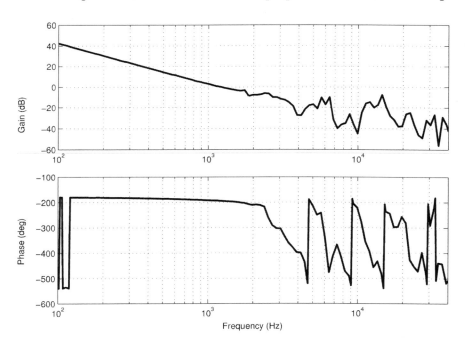

FIGURE 10.4
Frequency response of transfer function from u_v to y using PZT active suspension as a sensor with high frequency inner loop compensation. The VCM's and PZT active suspension's flexible resonant modes at 3.9, 6.0, 11.5, and 21.0 kHz are effectively damped with reduced relative degree. (From Ref. [17].)

High-Frequency Vibration Control

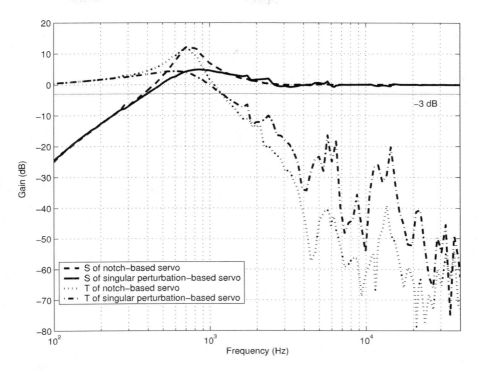

FIGURE 10.5
Frequency response of sensitivity transfer functions with the singular perturbation-based servo and conventional notch-based servo. (From Ref. [17].)

vibration suppression capabilities with a higher bandwidth and lower sensitivity function such that high-frequency disturbances would not be amplified by the control loop [19], which results in a lower position error.

10.2.4.2 Time Responses

For comparison purposes, the gain crossover frequency of both conventional notch-based control and the proposed control scheme with the PZT as an additional sensor are set at 700 Hz. The experimental step responses of 0.1 μm are shown in Figures 10.6 and 10.7, respectively. The slow control signal $\bar{u}_V$ and fast control signal $\tilde{u}_V$ are shown in Figure 10.8. It is noticed that the induced oscillations using the proposed control scheme are highly suppressed by the fast control signal $\tilde{u}_V$. A smaller overshoot is seen due to active vibration control, but a slower seek and settling time is a trade-off. However, it should be noted that a much higher gain crossover frequency can be designed as the first in-plane resonant mode is well damped, which will result in faster seek and settling times. The details about the discussion of the simulation and experimental results are seen in [17].

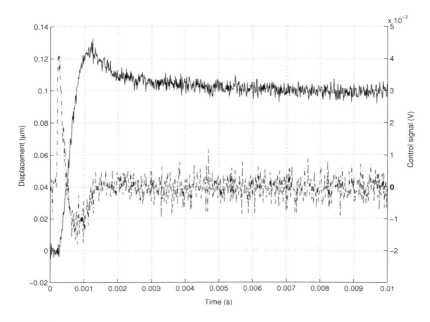

FIGURE 10.6
Experimental step response using conventional notch filter-based servo. Solid: displacement measured at tip of PZT active suspension. Dash-dot: control signal. (From Ref. [17].)

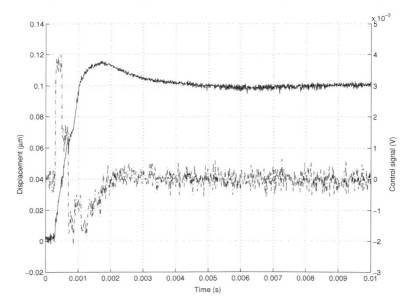

FIGURE 10.7
Experimental step response using the singular perturbation-based control. Solid: displacement measured at tip of PZT active suspension. Dash-dot: control signal. (From Ref. [17].)

High-Frequency Vibration Control

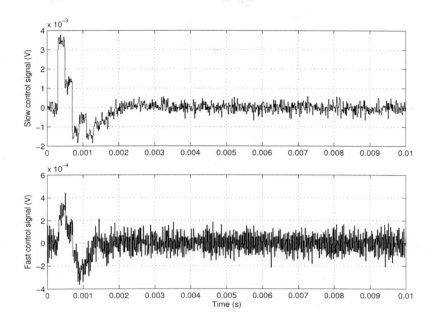

FIGURE 10.8
Control signals using the singular perturbation-based control. Top: slow control signal $\bar{u}_V$. Bottom: fast control signal $\tilde{u}_V$. (From Ref. [17].)

10.3 H_2 Controller Design

Consider the plant composed of a rigid part and several flexible modes and expressed as

$$P(s) = P_0(s) + \sum_{i=1}^{n} \frac{b_{1i}s + b_{0i}}{s^2 + 2\xi_i\omega_i s + \omega_i^2}, \qquad (10.3)$$

where $P_0(s)$ stands for the rigid part and the other term represents n flexible modes at the frequencies $\omega_i = 2\pi f_i$ (i = 1, 2, 3, ..., n). A basic feedback control loop for the positioning control of the plant is shown in Figure 10.9 (or Figure 1.5), where $C(z)$ is a feedback controller, d refers to a vibration that tends to excite the plant resonances, e denotes positioning error, and the reference r is assumed to be zero.

Figure 10.10 shows the schematic diagram of the active damping control. A sensor denoted by Γ is available to detect certain resonances of the plant with a sensing noise n_s. The control signal u_d from $C_d(z)$ to be designed is added to the feedback control signal u_c. It is seen that the active

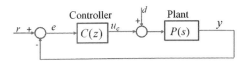

FIGURE 10.9
Block diagram of a feedback control loop.

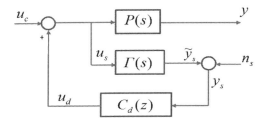

FIGURE 10.10
Block diagram of active damping for mechanical vibration control with sensor $\Gamma(s)$.

damping control loop is actually a feedback control and its stability must be considered.

Denote the new transfer function from y to u_c as T_{yu_c} with the sensor loop. The controller $C_d(z)$ is to be designed so that the resonance magnitude of T_{yu_c} is lowered as compared to $P(s)$. Thereby, the mechanical vibration due to these resonances will be reduced in the positioning error.

The sensor transfer function denoted by $\Gamma(s)$ can be written as

$$\Gamma(s) = \Gamma_n(s) + \Gamma_u(s) \tag{10.4}$$

where

$$\Gamma_n(s) = \sum_{i=1}^{n_n} \frac{\tilde{b}_{1i}s + \tilde{b}_{0i}}{s^2 + 2\tilde{\xi}_i\tilde{\omega}_i s + \tilde{\omega}_i^2} \tag{10.5}$$

includes the resonances corresponding to some of those in (10.3) and to be damped via the design of $C_d(z)$, and

$$\Gamma_u(s) = \sum_{i=n_n+1}^{n_n+n_u} \frac{\tilde{b}_{1i}s + \tilde{b}_{0i}}{s^2 + 2\tilde{\xi}_i\tilde{\omega}_i s + \tilde{\omega}_i^2} \tag{10.6}$$

includes the resonances detected by the sensor, but not necessarily to be damped because they are not excited significantly and/or not able to be damped due to the limitation of the sampling rate for the sensor signal.

High-Frequency Vibration Control

This portion of the sensor model as well as its variation is considered as uncertainty, and $\Gamma_n(s)$ is considered as a nominal model. This uncertainty is expressed as multiplicative uncertainty and defined by

$$\Delta(\omega) = \max_{i=1}^{p} \left| \frac{\Gamma_i(j\omega) - \Gamma_n(j\omega)}{\Gamma_n(j\omega)} \right| \qquad (10.7)$$

where p is the number of measurements and $\Gamma_i(j\omega)$ is the actual frequency response of the sensor in the i-th measurement, and $\Gamma_n(j\omega)$ is the frequency response of the nominal model in (10.5). An approximate bounding transfer function $W(s)$ can be found so that

$$|W(j\omega)| \geq \Delta(\omega). \qquad (10.8)$$

10.4 Design of $C_d(z)$ with H_2 Method and Notch Filters

The resonance components in (10.5) will be used to control the resonance vibration via the implementation of $C_d(z)$. The unwanted resonances in (10.6) must also be treated through $C_d(z)$ in order to make the loop with the sensor stable. Notch filters can be used to compensate for these unwanted resonances modeled as in (10.6).

The task of active damping is to lower the resonance magnitude of T_{yu_c}, which implies that the sensor output amplitude will be reduced by the active damping. The design of $C_d(z)$ is then formulated as an H_2 control problem. That is, to design a controller $C_d(z)$ such that the H_2 norm of the transfer function from $w_2 = [u_c \; n_s]^T$ to the sensor output $\tilde{y}_s$, denoted by $\|T_{\tilde{y}_s w_2}\|_2$, is minimized. The new system T_{yu_c} will have the resonance magnitude lowered as compared to the original plant $P(s)$.

The controller design is conducted in discrete-time domain. $\Gamma_n(z)$ is the discretized form of $\Gamma_n(s)$ and its state-space description is (A_s, B_s, C_s, D_s) with state $x_s(k)$. We consider the following system:

$$x(k+1) = Ax(k) + B_1 w_2(k) + B_2 u_d(k) \qquad (10.9)$$

$$y_s(k) = C_1 x(k) + D_{11} w_2(k) + D_{12} u_d(k) \qquad (10.10)$$

$$\tilde{y}_s(k) = C_2 x(k) + D_{21} w_2(k) + D_{22} u_d(k) \qquad (10.11)$$

where

$$x = x_s, \quad A = A_s, \quad B_1 = \begin{bmatrix} B_s & 0 \end{bmatrix}, \quad B_2 = B_s \qquad (10.12)$$

$$C_1 = C_s, \quad D_{11} = \begin{bmatrix} D_s & 1 \end{bmatrix}, \quad D_{12} = D_s \qquad (10.13)$$

$$C_2 = C_s, \quad D_{21} = \begin{bmatrix} D_s & 0 \end{bmatrix}, \quad D_{22} = D_s. \qquad (10.14)$$

An H_2 controller in the form of (A_c, B_c, C_c, D_c) from y_s to u_d such that $\left\| T_{\tilde{y}_s w_2} \right\|_2^2 < \mu$ can be obtained by using the following parameterization approach [20]:

$$D_c = R, C_c = (L - RC_1 X)\Lambda^{-1} \qquad (10.15)$$

$$B_c = \Xi^{-1}(U - YB_2 R) \qquad (10.16)$$

$$A_c = \Xi^{-1}[Q - Y(A + B_2 D_c C_1)X - \Xi B_c C_1 X - YB_2 C_c \Lambda]\Lambda^{-1} \qquad (10.17)$$

where Ξ and Λ are nonsingular such that $\Xi\Lambda = S - YX$, the matrices X, L, Y, Q, R, J, S and the symmetric matrices P_2, H_2, and W_2 are the variables of the following LMIs:

$$\text{trace}\{W_2\} \le \mu$$

$$\begin{bmatrix} W_2 & C_2 X + D_{22} L & C_2 + D_{22} RC_1 \\ * & X + X^T - P_2 & I + S^T - J \\ * & * & Y + Y^T - H_2 \end{bmatrix} > 0 \qquad (10.18)$$

$$\begin{bmatrix} P_2 & J & AX + B_2 L & A + B_2 RC_1 & A + B_2 RD_{11} \\ * & H_2 & Q & YA + UC_1 & YB_1 + UD_{11} \\ * & * & X + X^T - P_2 & I + S^T - J & 0 \\ * & * & * & Y + Y^T - H_2 & 0 \\ * & * & * & * & I \end{bmatrix} > 0 \quad (10.19)$$

where I is the identity matrix and $*$ denotes an entry that can be deduced from the symmetry of the LMIs. The minimization of the H_2 norm square $\left\| T_{\tilde{y}_s w_2} \right\|_2^2$ is equivalent to the minimization of trace$\{W_2\}$ subject to the LMIs (10.18) and (10.19).

10.5 Design of Mixed H_2/H_∞ Controller $C_d(z)$

In the previous section, notch filters are used to compensate for $\Gamma_u(s)$ in order to make sure that the sensor loop is stable. In this section, it is considered as

High-Frequency Vibration Control

uncertainty, and a mixed H_2/H_∞ method is used to make sure that the sensor loop is stable simultaneously when using the H_2 scheme to damp the system resonances. Specifically, we need to ensure the system stability against the unmodeled high-frequency dynamics of the sensor, i.e., the constraint $\|TW\|_\infty < 1$ is to be met, where T is the closed-loop transfer function and W is the bounding function of the unmodeled dynamics which was derived earlier. Therefore, we have the mixed H_2/H_∞ control scheme as shown in Figure 10.11, where $w \in l_2[0,\infty)$, as a disturbance input or a reference is used to generate the closed-loop T for the H_∞ constraint. Clearly, the transfer function from w to z is TW, where

$$T = \frac{\Gamma C_d}{1 - \Gamma C_d}. \tag{10.20}$$

Let (A_w, B_w, C_w, D_w) with state $x_w(k)$ be the state-space model of $W(z)$, which is the discretized form of $W(s)$. The state-space representation for the system in Figure 10.11 is derived as follows:

$$x(k+1) = Ax(k) + B_1 w_2(k) + B_i w(k) + B_2 u_d(k) \tag{10.21}$$

$$y_s(k) = C_1 x(k) + D_{11} w_2(k) + D_{1i} w(k) + D_{12} u_d(k) \tag{10.22}$$

$$\tilde{y}_s(k) = C_2 x(k) + D_{21} w_2(k) + D_{22} u_d(k) \tag{10.23}$$

$$z(k) = C_i x(k) + D_{i1} w(k) + D_{i2} u_d(k) \tag{10.24}$$

where

$$x = \begin{bmatrix} x_s^T & x_w^T \end{bmatrix}^T \tag{10.25}$$

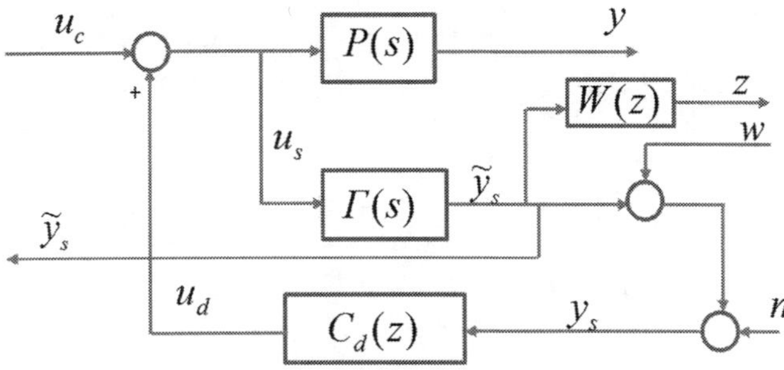

FIGURE 10.11
Block diagram of mixed H_2/H_∞ active damping for mechanical vibration control with sensor $\Gamma(s)$.

$$A = \begin{bmatrix} A_s & 0 \\ B_w C_s & A_w \end{bmatrix}, \quad B_1 = \begin{bmatrix} B_s & 0 \\ B_w D_s & 0 \end{bmatrix}, \quad B_i = 0, \quad B_2 = \begin{bmatrix} B_s \\ B_w D_s \end{bmatrix} \quad (10.26)$$

$$C_1 = [\; C_s \quad 0 \;], \quad D_{11} = \begin{bmatrix} D_s & 1 \end{bmatrix}, \quad D_{1i} = 1, \quad D_{12} = D_s \quad (10.27)$$

$$C_2 = C_s, \quad D_{21} = \begin{bmatrix} D_s & 0 \end{bmatrix}, \quad D_{22} = D_s \quad (10.28)$$

$$C_i = \begin{bmatrix} C_s & C_w \end{bmatrix}, \quad D_{i1} = 0, \quad D_{i2} = D_w D_s. \quad (10.29)$$

To solve the controller as in (10.15)–(10.17), the following LMI is needed in addition to (10.18) and (10.19) [20].

$$\begin{bmatrix} P_i & J_i & AX + B_2 L & A + B_2 R C_1 & B_i + B_2 R D_{1i} & 0 \\ * & H_i & Q & YA + U C_1 & YB_i + U D_{1i} & 0 \\ * & * & X + X^T - P_i & I + S^T - J & 0 & X^T C_i^T + L^T D_{i2}^T \\ * & * & * & Y + Y^T - H & 0 & C_1^T R^T D_{i2}^T + C_i^T \\ * & * & * & * & I & D_{1i}^T R^T D_{i2}^T + D_{i1}^T \\ * & * & * & * & 0 & \gamma I \end{bmatrix} > 0$$

$$(10.30)$$

where P_i, J_i, and H_i are variables and the variables P_i and H_i are symmetric matrices.

Considering the fact that the sensor picking up unwanted resonance modes is difficult to be avoided due to the limitation of sensor location, the proposed mixed H_2/H_∞ method for active damping control is appropriate from a practical point of view. And, the controller design is easily conducted using this LMI approach.

10.6 Application Results

10.6.1 System Modeling

We now consider the vibration control for a VCM actuation system. As shown in the frequency responses in Figure 10.12, there are six resonances at 6, 8.5, 10, 16, 21, and 28 kHz.

High-Frequency Vibration Control

A PZT sensor is available to detect some of these resonances, and the frequency responses of its transfer function $\Gamma(s)$ are measured and shown in Figure 10.13. The active damping control $C_d(z)$ is to be designed using the proposed methods and 60 kHz sampling rate. The controller $C(z)$ at 30 kHz sampling rate is a feedback controller to meet the basic control system requirements such as a certain bandwidth and stability margins; its design is omitted here.

According to Figure 10.13, the sensor has detected the modes at 6, 8.5, 10, 16, 20, and 28 kHz, etc. In this situation, the sensed modes at 6, 8.5, and 10 kHz will be used to control the corresponding resonances of the VCM plant, and others are considered as uncertainty. Therefore, in (10.4), 6, 8.5, and 10 kHz modes are used to create $\Gamma_n(s)$, and others are included in $\Gamma_u(s)$. Figure 10.13 shows the frequency responses of $\Gamma_n(s)$ as compared to the measured ones of $\Gamma(s)$.

The uncertainty is then obtained with $\Gamma_n(s)$ and several measurements based on (10.7). Figure 10.14 shows the uncertainty and the magnitude of its bounding function $W(s)$.

10.6.2 H_2 Active Damping Control

In this section, the proposed H_2 method applies to $\Gamma_n(s)$ for the control of the plant resonances at 6, 8.5, and 10 kHz. The sensing noise is assumed to occupy 10% of the sensor signal. Three notch filters are needed to compensate for 16, 20, and 28 kHz modes of the sensor in order to have a stable

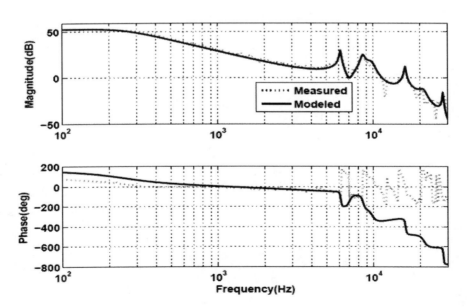

FIGURE 10.12
Frequency responses of a VCM actuator.

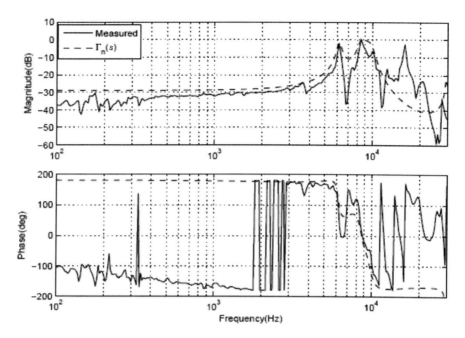

FIGURE 10.13
Measured frequency responses of the sensor transfer function and $\Gamma_n(s)$.

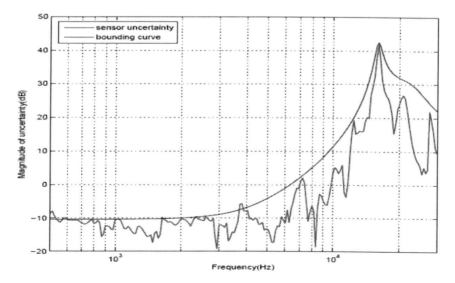

FIGURE 10.14
Magnitude of uncertainty.

High-Frequency Vibration Control

control loop with the sensor. The resultant damped system T_{yu_c} is shown in Figure 10.15 with comparison to the VCM actuator system $P(s)$. It is seen that the magnitudes of the resonances at 6, 8.5, and 10 kHz have all been lowered.

Here, the notch filters are used to compensate for the unwanted resonance modes. However, the notch filter will degrade the system performance by inducing addition phase delay, and it is not able to guarantee the loop stability once the resonances shift. Moreover, the notch filter method usually causes a high-order controller as each resonance requires a notch filter. In view of this, the mixed H_2/H_∞ method is necessary.

10.6.3 Mixed H_2/H_∞ Active Damping Control

$\Gamma_n(s)$ involves the modes at 6, 8.5, and 10 kHz which are to be damped, as shown in Figure 10.15 with comparison to the measured sensor transfer function $\Gamma(s)$. Other modes are considered as uncertainty and covered by the uncertainty bounding function $W(s)$, as seen in Figure 10.14.

With the consideration of the uncertainty $W(s)$, a controller $C_d(z)$ is designed by using the proposed mixed H_2/H_∞ method. The designed controller is able to guarantee the system stability with the presence of the high-frequency resonances detected by the sensor but not used for the active damping. Figure 10.16 shows the simulated frequency responses of T_{yu_c} with the designed $C_d(z)$ as compared with the VCM model $P(s)$. It is observed in Figure 10.16 that the mixed H_2/H_∞ controller also achieves the damping effect for the resonances at 6, 8.5, and 10 kHz.

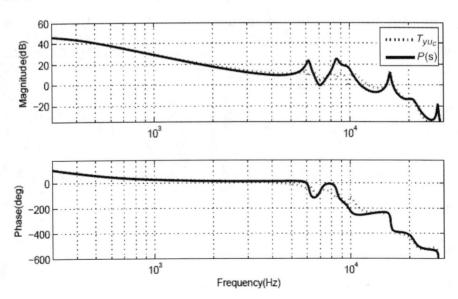

FIGURE 10.15
Frequency responses of T_{yu_c} as compared with the VCM plant $P(s)$ (H_2 method).

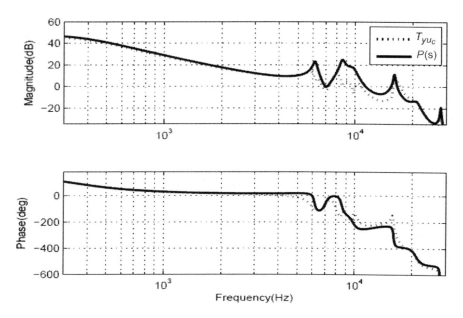

FIGURE 10.16
Frequency responses of T_{yu_c} as compared with the VCM plant (mixed H_2/H_∞ method).

10.6.4 Experimental Results

Experiment was carried out on a control testing platform built with the equipment of LDV and dSPACE. Figure 10.17 shows the measured frequency responses of T_{yu_c} with and without $C_d(z)$. The magnitude from 6 to 10 kHz is lowered due to the controller $C_d(z)$. The experimental testing also shows that the closed loop with the feedback controller $C(z)$ remains stable when the controller $C_d(z)$ is enabled. As a white noise used as the disturbance d excites the dynamics of the VCM actuator, the error e is investigated and its spectrum is plotted in Figures 10.18 and 10.19, where it is observed that the error is much smaller in the frequency range from 5 to 11 kHz due to the active damping controller. It is evaluated that about 60% improvement is achieved for both methods of H_2 and mixed H_2/H_∞. Therefore, it is concluded that the controllers designed with the two methods are both practically implementable and have achieved the effectiveness of vibration control as expected.

10.7 Conclusion

The PZT elements in the dual-stage actuation systems are used as additional sensors to enable high-frequency vibration control. The controller design

High-Frequency Vibration Control

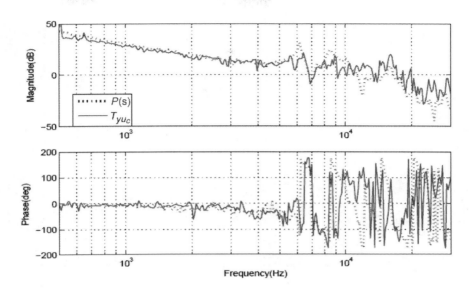

FIGURE 10.17
Measured frequency responses of T_{yu_c} as compared with the VCM plant $P(s)$.

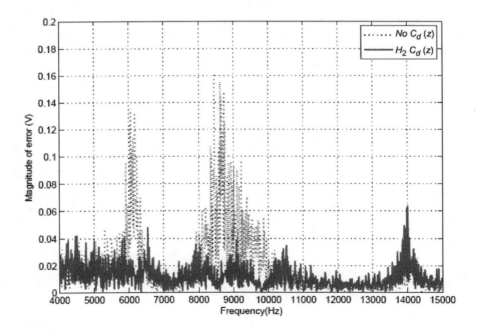

FIGURE 10.18
Error spectrum with and without $C_d(z)$ designed with the H_2 method.

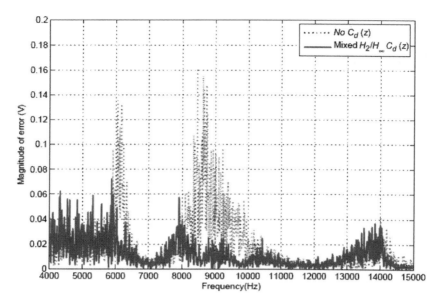

FIGURE 10.19
Error spectrum with and without $C_d(z)$ designed with the mixed H_2/H_∞ method.

method based on the singular perturbation theory has been presented using the PZT elements as a fast subsystem observer to detect the structure's high-frequency dynamics. The slow and fast controllers are designed independently, parameterized by a single parameter ε for time scale separation.

The high-frequency vibration control problem has also been formulated as H_2 control problem and mixed H_2/H_∞ control problem and solved via LMI approaches. The resonance modes at 6, 8.5, and 10 kHz have been reduced in magnitude with the damping controllers. Reflected in the position error, the impact of these resonance modes induced vibration have been apparently reduced. With the resonance modes detected by the sensor but not to be damped considered as uncertainties, the proposed mixed H_2/H_∞ method is able to ensure the system stability when conducting the structural vibration control by the active damping. Thus, it is a more systematic method and is applicable in practice as well. In the next chapter, the dual-stage mechanism with self-sensing actuation, i.e., using the PZT as sensor and actuator simultaneously, will be studied.

References

1. H. W. Bode, *Network Analysis and Feedback Amplifier Design*, Princeton, Van Nostrand, 1945.

High-Frequency Vibration Control 153

2. B. Wu and E. Jonckheere, A simplified approach to Bode's theorem for continuous-time and discrete-time systems, *IEEE Trans. Automat. Contr.*, 37(11), pp. 1797–1802, 1992.
3. D. Abramovitch, T. Hurst, and D. Henze, The PES Pareto method: uncovering the strata of position error signals, *Proc. of the 1997 American Control Conf.*, Albuquerque, NM, June, 1997.
4. C.-I. Kang and A.-H. Kim, An adaptive notch filter for suppressing mechanical resonance in high track density disk drives, *Microsyst. Technol.*, 11(8), pp. 638–652, 2005.
5. G. F. Franklin, J. D. Powell, and A. Emaminaeini, *Feedback Control of Dynamic Systems*, 3rd Edition, Addison-Wesley, New York, 1994.
6. Y. Yamaguchi, K. Takahashi, and H. Fujita, Flow induced vibration of magnetic head suspension in hard disk drive, *IEEE Trans. Magn.*, 22(5), pp. 1022–1024, 1986.
7. H. Shimizu, T. Shimizu, M. Tokuyama, H. Masuda, and S. Nakamura, Numerical simulation of positioning error caused by air-flow-induced vibration of head gimbals assembly in hard disk drive, *IEEE Trans. Magn.*, 39(2), pp. 806–811, 2003.
8. Y. Huang, M. Banther, P. D. Mathur, and W. C. Messner, Design and analysis of a high bandwidth disk drive servo system using an instrumented suspension, *IEEE/ASME Trans. Mechatron.*, 4(2), pp. 196–206, 1999.
9. F. Y. Huang, T. Semba, W. Imaino, and F. Lee, Active damping in HDD actuator, *IEEE Trans. Magn.*, 37(2), pp. 847–849, 2001.
10. S.-H. Lee, C. C. Chung, and C. W. Lee, Active high-frequency vibration rejection in hard disk drives, *IEEE/ASME Trans. Mechatron.*, 11(3), pp. 339–345, 2006.
11. M. Kobayashi, S. Nakagawa, T. Atsumi, and T. Yamaguchi, High-bandwidth servo control designs for magnetic disk drives, 2001 *IEEE/ASME International Conf. on Advanced Intelligent Mechatronics Proceedings*, Como, Italy, July 8–12, 2001, pp. 1124–1129.
12. Y. Li, F. Marcassa, R. Horowitz, R. Oboe, and R. Evans, Track-following control with active vibration damping of a PZT-actuated suspension dual-stage servo system, *Proc. of the American Control Conf.*, Denver, Colorado, June 4–6, 2003, pp. 2553–2559.
13. X. Huang, R. Horowitz, and Y. Li, A comparative study of MEMS microactuator for use in a dual-stage servo with an instrumented suspension, *IEEE/ASME Trans. Mechatron.*, 11(5), pp. 524–532, 2006.
14. P. V. Kokotovic, H. K. Khalil, and J. O'Reilly, *Singular Perturbation Methods in Control: Analysis and Design*, Academic Press, London, 1986.
15. M. Hirata, T. Atsumi, A. Murase, and K. Nonami, Following control of a hard disk drive by using sampled-data H_∞ control, *Proceedings of the IEEE International CCA*, Kohala Coast-Island of Hawaii, HI, USA, August 22–26, 1999, pp. 182–186.
16. C. K. Pang, G. Guo, B. M. Chen, and T. H. Lee, Self-sensing actuation for nanopositioning and active-mode damping in dual-stage HDDs, *IEEE/ASME Trans. Mechatron.*, 11(3), pp. 328–338, 2006.
17. C. K. Pang, F. L. Lewis, S. S. Ge, G. Guo, B. M. Chen, and T. H. Lee, Singular perturbation control for vibration rejection in HDDs using the PZT active suspension as fast subsystem observer, *IEEE Trans. Ind. Electron.*, 54(3), pp. 1375–1386, 2007.
18. F. L. Lewis, S. Jagannathan, and A. Yecsildirek, *Neural Network Control of Robot Manipulators and Nonlinear Systems*, Taylor and Francis, London, 1999.

19. C. K. Pang, D. Wu, G. Guo, T. C. Chong, and Y. Wang, Suppressing sensitivity hump in HDD dual-stage servo systems, *Microsyst. Technol.*, 11(8–10), pp. 653–662, 2005.
20. M. C. de Oliveira, J. C. Geromel, and J. Bernussou, An LMI optimization approach to multiobjective and robust H_∞ controller design for discrete-time systems, *Proc. of the 38th IEEE Conf. on Decision and Control*, vol. 4, 1999, pp. 3611–3616.

11
Self-Sensing Actuation of Dual-Stage Systems

11.1 Introduction

In this chapter, the PZT (Pb-Zr-Ti) elements in the PZT secondary actuator discussed in Chapter 1 are used as a secondary actuator and a displacement sensor simultaneously with self-sensing actuation (SSA). The used displacement estimation circuit produces an estimated PZT secondary actuator's displacement with high SNR (signal-to-noise ratio) and nanometer resolution. The estimated displacement is used to decouple the dual-stage system control into two distinct loops for individual sensitivity function optimization as well as to design a robust controller to damp the mechanical structure high-frequency modes such as torsion mode and sway mode.

The SSA dual-stage control topology is shown in Figure 11.1 [1]. Unlike the conventional control structure where only measured sensor signal is used, this SSA control topology uses the measured sensor signal and the estimated PZT secondary actuator's displacement obtained from the SSA for feedback control. A bridge circuit is needed to obtain a real-time estimated displacement for actively decoupling the dual-stage loop so that the controller

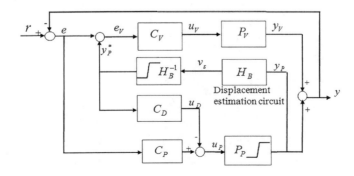

FIGURE 11.1
SSA dual-stage control topology. The circuit H_B allows an online estimation of the PZT secondary actuator's displacement y_P^*. (From Ref. [1].)

156   *Multi-Stage Actuation Systems and Control*

designs for the VCM (voice coil motor) and the PZT secondary actuator can be carried out independently. The estimated secondary actuator's displacement is also used to design a controller to damp the mechanical structure high-frequency modes via an inner loop compensation.

11.2 Estimation of PZT Secondary Actuator's Displacement y_P^*

In this section, how to obtain an online estimation of PZT secondary actuator's displacement y_P^* will be addressed.

11.2.1 Self-Sensing Actuation and Bridge Circuit

The piezoelectric material can be used as an actuator and sensor simultaneously, or commonly known as SSA, which is attractive in active control applications because the actuator and the sensor arrangement is a truly collocated (or in many applications near collocated) pair, hence avoiding non-minimum phase zero dynamics which will degrade control performance. In the PZT secondary actuators, the PZT elements can be modeled as a capacitance in series with a variable voltage source, where the capacitor represents the dipoles and the variable voltage source represents the electric field setup by the dipoles during actuation. If the capacitance of the PZT elements is known, the variable voltage can be decoupled using a bridge circuit, and hence strain/displacement information is available. The bridge circuit shown in Figure 11.2 [2] is used to generate displacement (proportional to strain) required for active control. In the bridge circuit, the voltage v_P generated is proportional to y_P of the PZT secondary actuator and is decoupled from the control voltage u_P using a differential amplifier. From Figure 11.2, we have

$$v_1 = \frac{C_{\mathrm{PZT}}}{C_1 + C_{\mathrm{PZT}}}(u_P - v_P) \tag{11.1}$$

$$v_2 = \frac{C_2}{C_2 + C_3} u_P \tag{11.2}$$

$$v_s = v_2 - v_1$$
$$= \left(\frac{C_{\mathrm{PZT}}}{C_1 + C_{\mathrm{PZT}}} - \frac{C_2}{C_2 + C_3} \right) \cdot u_P + \frac{C_{\mathrm{PZT}}}{C_1 + C_{\mathrm{PZT}}} \cdot v_P. \tag{11.3}$$

C_{PZT} is the measured capacitance of the PZT element. Resistors R_1 and R_3 are placed in parallel with the capacitors C_1 and C_3 to prevent the DC (direct

Self-Sensing Actuation

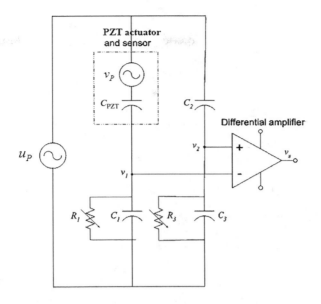

FIGURE 11.2
Bridge circuit employing the PZT secondary actuator as actuator and sensor simultaneously. The PZT elements are modeled as a capacitor C_{PZT} with dependent voltage source v_P. (From Ref. [1].)

current) drift. When $C_1 = C_2 = C_3 = C_{PZT}$, the bridge circuit is balanced and a PZT displacement estimator is established with

$$v_s = \frac{C_{PZT}}{C_1 + C_{PZT}} v_P = \frac{v_P}{2}. \tag{11.4}$$

As such, v_P is decoupled from the control signal u_P and can be used for control purpose.

Applying SSA in dual-stage actuation systems only requires additional cheap electronic circuitry and does not reduce the effective actuation of the PZT actuator. The only trade-off is that a larger control signal u_P is needed for the same amount of displacement actuation as compared to the case with no bridge circuit.

11.2.2 PZT Displacement Estimation Circuit H_B

The PZT displacement estimation circuit H_B consists of the PZT secondary actuator with sensing function enabled and the differential amplifier setup. H_B is modeled by a transfer function $H_B(s) = v_s(s)/y_P(s)$, where v_P (hence v_s) arises from y_P. The frequency response of $H_B(s)$ is measured with swept sine excitation and is shown in Figure 11.3. And, the following sixth-order transfer function is used for $H_B(s)$.

$$H_B(s) = K_B \frac{\omega_1^2}{s^2 + 2\zeta_1\omega_1 s + \omega_1^2} \cdot \frac{\omega_2^2}{s^2 + 2\zeta_2\omega_2 s + \omega_2^2} \cdot \frac{\omega_3^2}{\tau} \cdot \frac{s+\tau}{s^2 + 2\zeta_3\omega_3 s + \omega_3^2}. \quad (11.5)$$

The values of K_B, ζ_i, ω_i ($i = 1, 2,$ and 3), and τ are shown in Table 11.1.

The inverse $H_B^{-1}(s)$ is constructed as

$$H_B^{-1}(s) = \frac{\tau \omega_\beta^5}{K_B \omega_1^2 \omega_2^2 \omega_3^2} \cdot \frac{s^2 + 2\zeta_1\omega_1 s + \omega_1^2}{(s+\tau)(s+\omega_\beta)} \cdot \frac{s^2 + 2\zeta_2\omega_2 s + \omega_2^2}{(s+\omega_\beta)^2} \cdot \frac{s^2 + 2\zeta_3\omega_3 s + \omega_3^2}{(s+\omega_\beta)^2}$$

(11.6)

where $\omega_\beta = \pi \frac{f_s}{\beta}$, and $1 < \beta < 1.2$ is chosen usually so that the $H_B^{-1}(s)$ is realizable. The poles are chosen near the Nyquist frequency $f_s/2$ to maximize the dynamic range of H_B^{-1}. A digital $H_B^{-1}(z)$ is then available from (11.6) at a sampling frequency f_s of 100 kHz to provide an estimate of the PZT

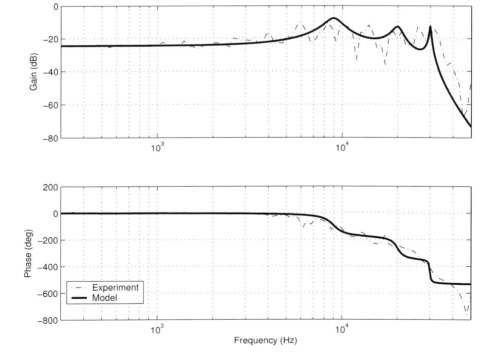

FIGURE 11.3
Frequency response of displacement estimation circuit H_B. (From Ref. [1].)

Self-Sensing Actuation

TABLE 11.1
Parameters of $H_B(s)$

K_B	0.06
ζ_1	0.2
ζ_2	0.09
ζ_3	0.03
ω_1	$2\pi\, 9 \times 10^3$
ω_2	$2\pi\, 20 \times 10^3$
ω_3	$2\pi\, 30 \times 10^3$
τ	$2\pi \times 10^3$

Source: From Ref. [1].

displacement y_P^*. A saturation function is also included to mimic the saturation of the PZT secondary actuator, and v_s is channeled into the digital inverse $H_B^{-1}(z)$ so that its output y_P^* estimates the PZT displacement y_P, as shown in Figure 11.1.

With the PZT secondary actuator being set to actuate at about ±8 nm in radial directions, the frequency responses of the measured PZT displacement y_P from the equipment LDV (laser Doppler vibrometer) to the estimated one

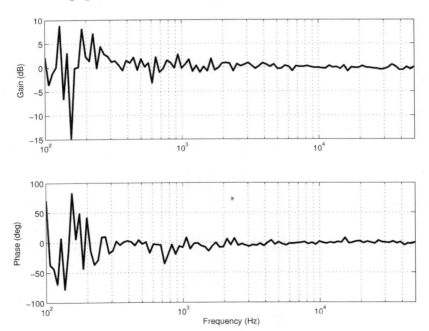

FIGURE 11.4
Frequency response of measured PZT secondary actuator's displacement from LDV y_P to estimated PZT displacement y_P^*. The PZT secondary actuator is set to actuate at about ±8 nm in radial directions. (From Ref. [1].)

y_P^* are measured and shown in Figure 11.4. It is seen that the estimated PZT displacement y_P^* correlates well with the measured displacement y_P from about 300 Hz onwards even at small displacements of ±8 nm. This makes the displacement estimation circuit H_B effective only in the high frequencies, which is tolerable as the error rejection in low-frequency range is mainly done by the VCM primary actuator. The high frequency accuracy is also essential for effective inner loop compensation via active control to actively damp the high-frequency vibration modes, which is to be discussed.

11.3 Design of Controllers

With the measured position error signal and the online estimated PZT secondary actuator's displacement y_P^*, we can proceed to design the controllers in Figure 11.1.

11.3.1 VCM Controller and Controller C_D

The VCM controller is designed with the traditional method introduced in "Remark 2.2".

SSA makes the collocation of PZT secondary actuator and displacement sensor nearly achieved. The estimated PZT displacement y_P^* can be used for many control methodologies, e.g., DISO (dual-input single-output) robust control synthesis. In this section, a robust controller design is presented for damping the high-frequency modes such as mechanical structure's torsion and sway modes.

The PZT secondary actuator behaves like a pure gain in low frequencies coupled with a number of resonance modes at high frequencies as seen in Figure 11.5. The PZT secondary actuator under consideration has identified the torsion modes at 4.31 and 6.52 kHz and the sway mode at 21.08 kHz. The identified $P_P(s)$ can be found in [1]. The controller $C_{D,i}(s)$ for each resonance mode i is proposed as

$$C_{D,i}(s) = K_{D,i} \cdot \frac{s + \dfrac{\omega_{n,i}}{\varepsilon_i}}{s + \dfrac{\kappa_i \omega_{n,i}}{\varepsilon_i}} \cdot \frac{s + \omega_{n,i}}{s + \dfrac{\varepsilon_i \omega_{n,i}}{\kappa_i}} \tag{11.7}$$

to increase the damping ratios or smaller magnitudes of the PZT secondary actuator's torsion modes and sway mode. $K_{D,i}$ is set to ε_i in most cases. ε_i and κ_i are tuning parameters. $1 < \varepsilon_i < q$ is chosen usually for robustness against natural frequency variations $q\%$, and $5 < \kappa_i < 15$ is chosen to determine the

Self-Sensing Actuation

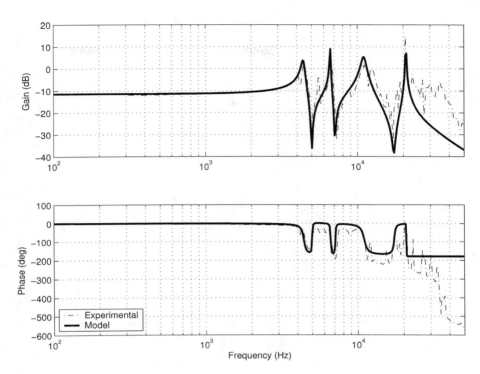

FIGURE 11.5
Frequency response of PZT secondary actuator. (From Ref. [1].)

amount of phase lift required to stabilize the resonance mode at the natural frequency $\omega_{n,i}$. The overall controller $C_D(s)$ is given by

$$C_D(s) = \prod_{i=1}^{N} C_{D,i}(s) \tag{11.8}$$

where N is the total number of resonant modes. The controller $C_D(s)$ in essence increases the gain of each resonant mode but stabilizes the PZT secondary actuator loop using a phase lead from the zeros at $(\omega_{n,i}/\varepsilon_i)$. $C_D(s)$ can be thought of as a general case of traditional PPF (positive position feedback) vibration controller [3] with additional zeros. The zeros prevent causality issues in PPF arrangements and improve the robust stability margin of the closed-loop system, as will be illustrated later.

11.3.2 PZT Controller

Let the transfer function of the PZT actuator with the controller C_D from $u_P(s)$ to $y_P(s)$ be P_P^C. The frequency responses from experiment and mathematical model P_P^C are plotted in Figure 11.6. The PZT controller of the form

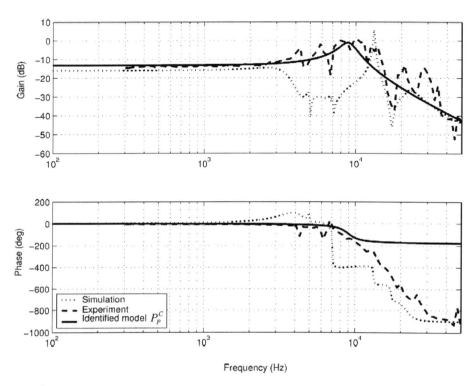

FIGURE 11.6
Identified model of PZT secondary actuator with the controller C_D. (From Ref. [1].)

$$C_P(s) = K_P \cdot \frac{\sigma^2}{(s+\sigma)^2} \cdot \frac{s + \dfrac{2\pi f_P}{\beta}}{s + 2\pi \beta f_P} \tag{11.9}$$

is used to control P_P^C. f_P is the gain crossover frequency of the PZT open-loop path and β is a tuning parameter for intended disturbance attenuation. $1 < \beta < 1.2$ is used typically, and K_P can be found with the relation $|C_P(j2\pi f_P)P_P(j2\pi f_P)| = 1$. σ is chosen to be $\sigma < 100$, for practical implementation purpose, as it prevents a large low frequency gain which saturates the PZT secondary actuator and causes integrator wind-up.

The PZT controller C_P in (11.9) is in essence a double integrator in series with a lead compensator. The integrators are essential to test the effectiveness of the bridge circuit in estimating the PZT actuator's displacement, and at the same time filter the high frequency noise in the displacement estimation circuit. With the integrators, the steady-state accuracy in the PZT actuator loop is ensured for fast tracking and error rejection. The lead compensator

Self-Sensing Actuation 163

is added to improve the phase margin at the gain crossover frequency and hence closed-loop stability.

11.4 Performance Evaluation

11.4.1 Effectiveness of C_D

Consider Figure 11.1. If the closed-loop system is stable, the following closed-loop equation holds

$$\frac{y_P(s)}{u_P(s)} = \frac{P_P(s)}{1 + C_D(s)P_P(s)}. \tag{11.10}$$

The gains of the modes are suppressed by the controller C_D at the frequencies $\omega_{n,i}$ of the resonant modes of the PZT secondary actuator P_P effectively by a factor of $\left| C_D\left(j\omega_{n,i} \right) \right|^{-1}$, as

$$\left| \frac{y_P(j\omega_{n,i})}{u_P(j\omega_{n,i})} \right| = \left| \frac{P_P(j\omega_{n,i})}{1 + C_D(j\omega_{n,i})P_P(j\omega_{n,i})} \right| \approx \frac{\left| P_P(j\omega_{n,i}) \right|}{\left| C_D(j\omega_{n,i})P_P(j\omega_{n,i}) \right|} = \frac{1}{\left| C_D(j\omega_{n,i}) \right|} \tag{11.11}$$

if $\left| C_D(j\omega_{n,i})P_P(j\omega_{n,i}) \right| \gg 1$.

The simulated frequency responses of the controller $C_D(s)$ and the open-loop transfer function $C_D(s)P_P(s)$ are shown in Figure 11.7. Figure 11.8 shows that the torsion modes at 4.31 kHz and 6.52 kHz as well as the sway mode at 21.08 kHz are all damped by the controller C_D. However, from Figure 11.8, it can be seen that the controller $C_D(s)$ is more effective in suppressing the sway mode of the PZT actuator at 21.08 kHz (>30 dB) than the torsion modes at 4.31 and 6.52 kHz (about 5 dB). This is because the estimated PZT displacement y_P^* from the displacement estimation circuit H_B is only effective in measuring in-plane sway modes of the PZT actuator and not the out-of-plane torsion modes. As another result, the simulated step responses of the PZT secondary actuator with and without the controller C_D are shown in Figure 11.9.

11.4.2 Position Errors

Additive white noise with zero mean and variance of 0.01 is added to the PZT displacement to mimic sensing noise in the PZT elements. And, the vibration and noise models in [4] are used. The 3σ values of the position error signals with different control schemes used in the PZT secondary actuator loop are shown in Figure 11.10.

164　　　　　　　　　　　　　　　　　　*Multi-Stage Actuation Systems and Control*

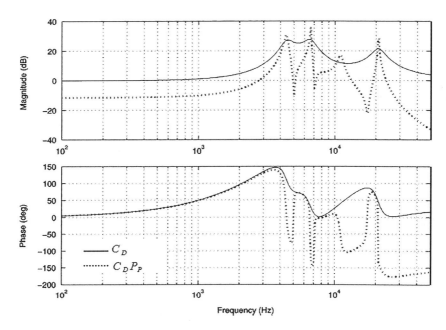

FIGURE 11.7
Simulated frequency responses of C_D and $C_D P_p$. (From Ref. [1].)

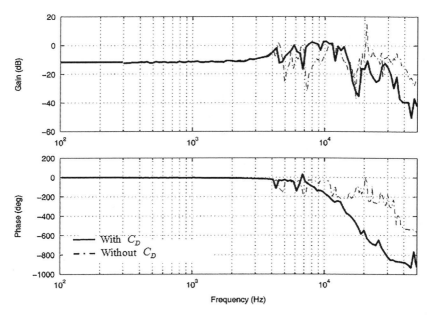

FIGURE 11.8
Experimental frequency responses of PZT secondary actuator. (From Ref. [1].)

Self-Sensing Actuation

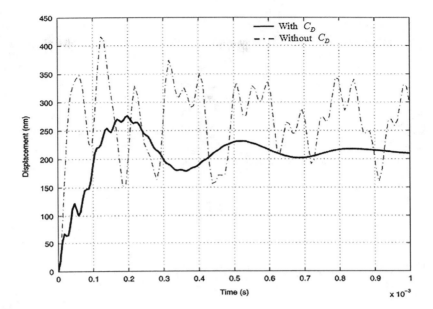

FIGURE 11.9
Simulated step responses of PZT secondary actuator with and without the controller C_D. (From Ref. [1].)

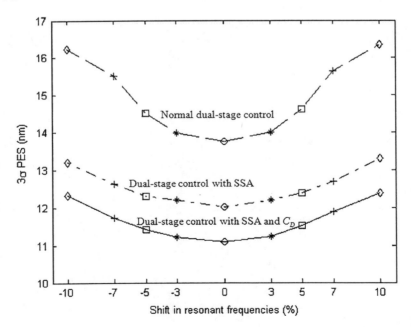

FIGURE 11.10
Comparison of 3σ value of position errors with different control schemes. (From Ref. [1].)

The conventional decoupled control scheme with the structure in Figure 2.1 uses digital notch filters to attenuate the gain of the resonant modes in the PZT secondary actuator. In the decoupled control scheme with SSA, digital notch filters together with $C_D = 0$ are used as one case shown in Figure 11.10.

The SSA control topology enables the controller C_D to damp the mechanical structure high-frequency modes such as torsion modes and sway modes. It can be seen that when SSA and C_D are employed simultaneously, better positioning capabilities are achieved with a reduction of up to 20% in the 3σ value of the position error signal, and the control scheme with C_D is robust for $\pm 10\%$ shifts in the frequencies of the resonant modes [1].

11.5 Conclusion

In this chapter, the SSA mechanism has been discussed for dual-stage actuation systems, and a cheap and collocated PZT actuator's displacement sensor has been constructed with nanoposition resolution. The PZT actuator's displacement is estimated by designing a circuit H_B with high correlation and high SNR without compromising the effective actuation of the PZT secondary actuator. The estimated displacement is then used to design a robust controller to actively damp the structure's torsion and sway modes as well as decouple the dual-stage loop. It has been shown that the designed dual-stage system control loop is effectively decoupled with SSA and uses the sensing signal from SSA to have significantly controlled high-frequency vibrations.

References

1. C. K. Pang, G. Guo, B. M. Chen, and T. H. Lee, Self-sensing actuation for nanopositioning and active-mode damping in dual-stage HDDs, *IEEE/ASME Trans. Mech.*, 11(3), pp. 328–337, 2006.
2. S. O. R. Moheimani, A survey of recent innovations in vibration damping and control using shunted piezoelectric transducers, *IEEE Trans. Control Syst. Technol.*, 11(4), pp. 482–494, 2003.
3. M. I. Friswell and D. J. Inman, The relationship between positive position feedback and output feedback controllers, *Smart Mater. Struct.*, 8, pp. 285–291, 1999.
4. C. Du, J. Zhang, and G. Guo, Vibration analysis and control design comparison of fluid bearing and ball bearing HDDs, *Proc. American Control Conf.*, Anchorage, AK, May 8–10, 2002, pp. 1380–1385.

12

Modeling and Control of a MEMS Micro X–Y Stage Media Platform

In this chapter, a micro X–Y stage with 6 mm × 6 mm recording media platform actuated by capacitive comb drives is introduced, which was once fabricated in a so-called "nanodrive" probe-based storage system (PBSS) [1–4]. The developed prototype was fabricated by micromachining techniques with the integration of a 40 nm thick polymethyl methacrylate (PMMA) recording media. Using finite element analysis (FEA) on the developed platform, the first two dominant in-plane resonant modes of the micro X–Y stage at 440 Hz are captured. A displacement range of up to 20 μm can be achieved at an input driving voltage of 55 V. A capacitive self-sensing actuation (CSSA) circuit and robust decoupling control scheme are also designed for the micro X–Y stage to have improved control performances. As an alternative futuristic data storage platform, the PBSS is successfully demonstrated in the "Millipede" project [1].

12.1 Introduction

With MEMS (microelectromechanical systems) technology becoming increasingly matured, micro X–Y stages have attracted many researchers, as it is frequently used in the development of small micro-fabricated motors and actuators. In [5], Choi et al. fabricated an electromagnetic micro X–Y stage to drive the recording media using bonding and assembling processes to fit the permanent magnets into their positions for alignments with the micro-coils. Kim et al. in [6] integrated a micro X–Y stage using silicon on glass substrate with a scanning range of 50 mm and moving area of the center platform for the recording media at about 20% of the total micro X–Y stage surface area. Alfaro et al. presented a micro media actuator for positioning the recording media in X–Y directions using silicon wafer double-side etching and wafer-to-wafer bonding processes for device fabrication in [7].

In this chapter, a novel design of the micro X–Y stage with a large movable media coated platform of dimensions 6 mm × 6 mm is presented for the nanopositioning and the control of the developed PBSS in the so-called "nanodrive". The proposed recording media area is about 36% of the total micro X–Y stage area. The fabrication process (including the integration of

PMMA recording media) is detailed and the prototype of the device was fabricated by micromachining techniques. Extending the SSA scheme in [8], a CSSA circuit is proposed for the MEMS comb drivers in the micro X–Y stage instead of the conventional thermal sensors used in [1], which is verified with experimental results. A robust decoupling control scheme based on open-loop H_∞ shaping [9] is also designed to reduce the mechanical crosstalk (axis coupling) in the micro X–Y stage during actuation.

The rest of this chapter is organized as follows. Section 12.2 describes the design, the fabrication process, and the FEA of the proposed micro X–Y stage. The CSSA for sensing the micro X–Y stage's displacement is designed in Section 12.3 and is verified with experimental results. Section 12.4 proposes a robust decoupling control method for high bandwidth and decoupled actuation of the micro X–Y stage. Finally, a conclusion is presented in Section 12.5.

12.2 MEMS Micro X–Y Stage

It is recalled that the components and fundamental operation principles of the "nanodrive" are addressed in Chapter 1. In this section, the design and fabrication of the MEMS micro X–Y stage for the "nanodrive" is detailed. The layout design of the micro X–Y stage is shown in Figure 12.1.

12.2.1 Design and Simulation of Micro X–Y Stage

The proposed micro X–Y stage [10] consists of a movable media platform, comb drive actuators, X and Y suspensions, springs, and stationary parts.

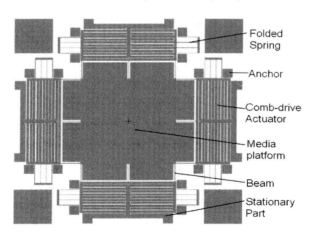

FIGURE 12.1
Simplified layout of a micro X–Y stage with 6 mm × 6 mm recording media. (From Ref. [10].)

MEMS Micro X–Y Stage Media Platform

The media platform is connected to the suspensions via supporting beams, and the suspensions are connected to the anchors through the folded springs. All anchors are fixed on the silicon-on-insulator (SOI) substrate with which the suspensions are electrically isolated from the stationary parts. With such a suspension design, a larger media area coupled with small mechanical crosstalk (axial coupling) will be achieved [11], which are critical concerns for nanoscale bit size and spacing in high storage capacities for smaller form factor PBSS. Also, four sets of comb drive actuators are used to drive the media platform moving in X and Y directions. In each actuator, comb drive interdigital capacitance electrodes attached to the arms of the suspensions and stationary parts are used to generate electrostatic actuation.

The electrostatic driving force F is expressed by

$$F = \frac{\varepsilon h n V^2}{g} \tag{12.1}$$

where ε is the permittivity of air, h is the height of the structure, n is the number of the electrode pairs, V is the driving voltage, and g is the gap width of the comb drive fingers.

The detailed parameters of the micro X–Y stage are given in Table 12.1.

12.2.1.1 Static

When a 55 V voltage is applied to one set of comb drive actuators, the overall electrostatic forces are calculated as 1 mN. Applying the electrostatic forces to the comb fingers for the X-axis actuation and running the static analysis in ANSYS, the displacement of the micro X–Y stage is determined and shown in Figure 12.2.

The displacement of the media platform achieved in the X-axis is about 20 μm. Correspondingly, the displacement in the Y-axis is 0.13 μm, which is very small as compared to the comb drive gap of 3 μm. This Y-axis displacement arises from the mechanical crosstalk (axial coupling) between the

TABLE 12.1

Design Parameters of Micro X–Y Stage

Over size	1 cm × 1 cm
Finger width	4 μm
Finger gap	3 μm
Comb-finger pairs	1,920
Spring width	8 μm
Height of spring/finger	60 μm
Spring length	1 mm
Media platform	6 mm × 6 mm

Source: From Ref. [10].

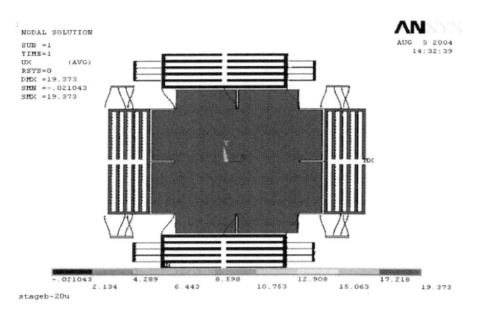

FIGURE 12.2
Displacement 20 μm of the recording media in X-axis under the driving voltage 55 V. (From Ref. [10].)

X- and Y-axes from the MEMS micro X–Y stage design. Upon a step response excitation of 55 V in the X-axis, other vibratory modes are excited and hence the maximum (worst case) displacement in the Y-axis of 0.13 μm is observed.

Ideally, the Y-axis should not be excited when the X-axis is actuated. However, in the proposed symmetrical structural design of the micro X–Y stage, there exists four groups of comb drives on the left, right, top, and down of the platform, as shown in Figure 12.1. When the platform is actuated to move in the X-axis, the spring beams of the top and down actuators will be arranged to move in a push-pull configuration. The bar supporting the spring (i.e., the top and down actuators) is not perfectly rigid, and the small deformation at that bar and the springs of top and down actuators actually cause the displacement in the Y-axis of about 0.6% as the forces required for actuation are merely provided by the comb fingers at the left side only. As such, the worst-case displacement decouple ratio y/x is 0.6% at the maximum driving voltage of 55 V.

12.2.1.2 Dynamic

To investigate the dynamic performance of the micro X–Y stage, modal and harmonic analysis are carried out using FEA in ANSYS. The first in-plane resonant mode of the micro X–Y stage at its first and second resonant frequencies of 440 Hz is shown in Figure 12.3.

MEMS Micro X–Y Stage Media Platform

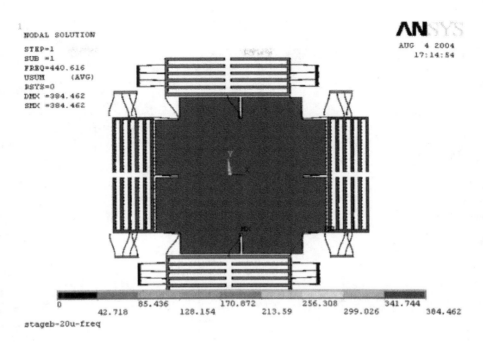

FIGURE 12.3
The first two resonant frequencies are at 440 Hz. The mode is an in-plane sway mode. (From Ref. [10].)

The frequency response of the media platform to an exciting force of 1 mN in X-axis is shown in Figure 12.4. It can be seen from Figure 12.4 that the micro X–Y stage exhibits a small "lumped" damping ratio and out-of-phase characteristics at the resonant frequency of 440 Hz and is perfectly symmetrical in actuator design.

12.2.2 Modeling of Micro X–Y Stage

Using the data points from FEA in ANSYS, the identified mathematical model $G_{xx}(s)$ in the X-axis of the MEMS micro X–Y stage with the first in-plane resonant frequency at 440 Hz is given by

$$G_{xx}(s) = 20 \cdot \frac{(2\pi 440)^2}{s^2 + 2 \cdot 0.0009 \cdot (2\pi 440)s + (2\pi 440)^2}, \quad (12.2)$$

which in essence is a second-order transfer function with a very small damping ratio and no high frequency uncertainties. This property is typical of MEMS devices.

Air damping is an important factor in the simulation of MEMS structures and plays a decisive role in comb driven micro resonators, especially at the

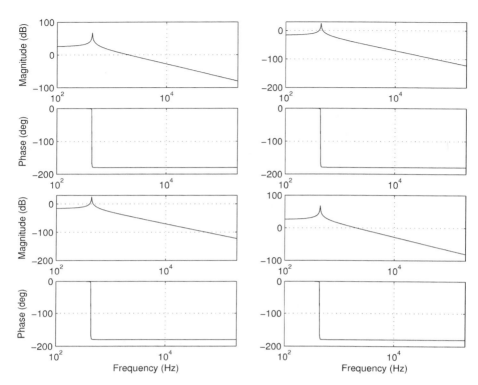

FIGURE 12.4
Frequency responses of the media platform to an exciting force 1 mN in X-axis. (From Ref. [10].)

high frequencies. However, it is extremely difficult to analyze the effects of air damping with theoretical relationships, e.g., using the revised Reynolds Equations with boundary conditions, slip, and gas rarefaction, etc., considerations for air shearing damping using ANSYS alone. As such, a lumped second-order resistor-inductor-capacitor electrical model or its mechanical counterpart, the mass-spring-damper (commonly used in the system identification of MEMS structures), is proposed to include the air film in the space between the comb fingers and the substrate or structure [12] to obtain (12.2).

Due to the large gain at the resonant frequency, the coupling effect between the main axes cannot be ignored. With the decoupling ratio between the main axes as 0.6%, the transfer functions of the MEMS micro X–Y stage in the Y-axis, $G_{yy}(s)$, as well as between the axes, $G_{xy}(s)$ and $G_{yx}(s)$, are

$$G_{yy}(s) = G_{xx}(s) \tag{12.3}$$

$$G_{xy}(s) = G_{yx}(s) = 0.006 G_{xx}(s). \tag{12.4}$$

MEMS Micro X–Y Stage Media Platform

As such, the transfer function $G(s)$ of the MEMS micro X–Y stage in both axes is

$$G(s) = \begin{bmatrix} G_{xx}(s) & G_{xy}(s) \\ G_{yx}(s) & G_{yy}(s) \end{bmatrix}$$

$$= 20 \cdot \frac{(2\pi 440)^2}{s^2 + 2 \cdot 0.0009 \cdot (2\pi 440)s + (2\pi 440)^2} \cdot \begin{bmatrix} 1 & 0.006 \\ 0.006 & 1 \end{bmatrix}.$$

(12.5)

It is worth noting that the dynamics of the micro stage in the Z-axis are unmodeled as the out-of-plane modes are uncontrollable. Although the existence of Z-axis flexible body modes affects the servo-positioning performance, they occur at low frequencies which are far from the open-loop gain crossover frequency to be detailed in future sections. Also, the spring constant in the Z-axis is much larger than that of the in-plane X and Y directions as the beam is a high aspect ratio structure of 60/8 = 7.5, and constraints have been included to fix the anchors in the Z-axis of the platform.

12.2.3 Fabrication of the MEMS Micro X–Y Stage

In practice, MEMS-based actuators are usually fabricated from a standard process including wet bulk etching, wafer bonding, surface micromachining, deep trench silicon micromachining, etc., similar to standard complementary metal oxide semiconductor (CMOS) processes. By aligning the actuators in a comb drive array, nanometer positioning accuracy with the electrostatic attractive forces from the MEMS actuators can be achieved.

An integrated fabrication procedure for the micro X–Y stage and ultra-thin recording media layer is shown in Figure 12.5. Starting from the SOI wafer, openings for releasing the large area media platform are etched until the buried oxide layer from the underside of the SOI. An ultra-thin layer of PMMA (which is commonly used as a recording media in PBSS) is formed on the topside by the spin-coat technique. The thickness of the PMMA film deposited is about 40 nm.

Next, a silicon dioxide layer is deposited by plasma enhanced chemical-vapor-deposited (PECVD) technique. The silicon dioxide layer is patterned as both a protective layer and a PMMA patterning mask. After PMMA patterned by oxygen plasma, a thick photoresist AZ4620 is spin-coated to cover the oxide protective layer and the patterned PMMA and is then used as a mask for X–Y stage etching process. The X–Y stage structures including comb drives, springs, beams, and suspensions are patterned by deep reactive ion etching (DRIE) and the photoresist is then removed by the oxygen plasma. Finally, the X–Y stage with the movable exposed PMMA layer platform is

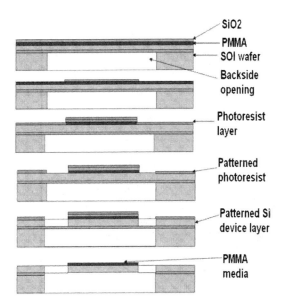

FIGURE 12.5
Fabrication process flow. (From Ref. [10].)

fully released after the silicon dioxide protective layer and SOI buried dioxide layer are removed by BOE (buffered oxide etching).

The partial view of the fabricated X–Y stage under scanning electron microscopy (SEM) is shown in Figure 12.6. The large area recording media platform is released from the SOI wafer substrate and is suspended by 12 supporting beams. All the movable parts are suspended by eight pairs of folded springs.

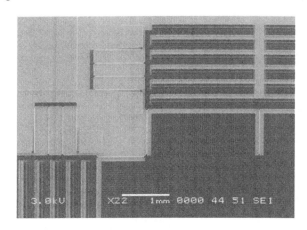

FIGURE 12.6
Partial view of the X–Y stage under SEM. (From Ref. [10].)

MEMS Micro X–Y Stage Media Platform

The details of the comb drives and the suspensions are shown in Figure 12.7. The finger width is 4 μm and the gap is 3 μm. The length of the finger is 75 μm, and the overlap of each pair of fingers is 30 μm.

The "H" structures are fabricated to protect the sidewalls of the folded springs and supporting beams during DRIE etching since the structure-area-density of the springs and beams is lower than that of the other areas, such as suspensions or comb drives, and is shown in Figure 12.8.

The undercut of the spring width and beam width is reduced, and the designed width of 8 μm is achieved.

FIGURE 12.7
Details of comb drives under SEM. Plan view of the fingers (top right). (From Ref. [10].)

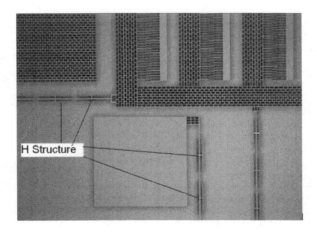

FIGURE 12.8
"H" structures for protecting sidewalls of springs during DRIE etching process. (From Ref. [10].)

12.3 Capacitive Self-Sensing Actuation

12.3.1 Design of CSSA Bridge Circuit

As the actuation of the MEMS micro X–Y stage is in the in-plane direction, non-intrusive online measurements of axial displacements and velocities are not possible even upon release of the MEMS micro X–Y stage from the wafer. As such, offline measurements from high-speed cameras coupled with image-processing techniques are commonly used for frequency response measurements.

Using the fact that the capacitance C_{MEMS} of the comb drives in the MEMS micro X–Y stage is proportional to the area of overlap A

$$C_{MEMS} = \frac{\varepsilon_r \varepsilon_0 A}{d} \quad (12.6)$$

which is in turn proportional to the displacement of the MEMS micro X–Y stage for the same comb width, a CSSA bridge circuit is proposed to achieve actuation and sensor capabilities simultaneously—similar to SSA in piezoelectric actuators [8]—to decouple the capacitance information, which is linear with the displacement of the MEMS micro X–Y stage.

The proposed CSSA scheme for the MEMS micro X–Y stage is shown in Figure 12.9.

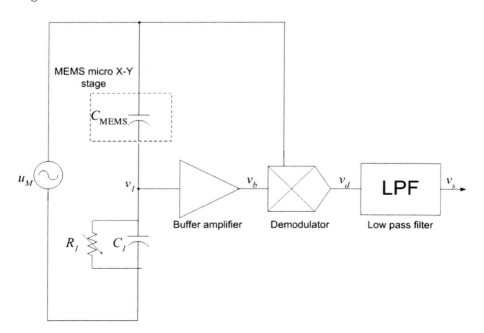

FIGURE 12.9
MEMS-based bridge circuit for CSSA. (From Ref. [10].)

MEMS Micro X–Y Stage Media Platform

In Figure 12.9, assume that the signal generator u_M produces a sinusoid $U \sin \omega t$ of angular frequency ω rad/s. The following equations hold

$$v_1 = \frac{C_{\text{MEMS}}}{C_1 + C_{\text{MEMS}}} U \sin \omega t \tag{12.7}$$

$$v_b = K_a v_1 \tag{12.8}$$

$$v_d = v_b u_M = K_a v_1 U \sin \omega t = \frac{U^2 K_a C_{\text{MEMS}}}{C_1 + C_{\text{MEMS}}} \sin^2 \omega t = \frac{U^2 K_a C_{\text{MEMS}}}{2(C_1 + C_{\text{MEMS}})} (1 - \cos 2\omega t) \tag{12.9}$$

where K_a is the gain of the buffer amplifier. After passing through the low-pass filter (LPF), the high-frequency sinusoid at 2ω is demodulated and v_s can be obtained as

$$v_s = \frac{U^2 K_a C_{\text{MEMS}}}{2(C_1 + C_{\text{MEMS}})} \tag{12.10}$$

and hence we can interpolate C_{MEMS} to be

$$C_{\text{MEMS}} = -\frac{2v_s C_1}{2v_s - U^2 K_a} \tag{12.11}$$

which is proportional to the area of overlap A (hence displacement) of the MEMS micro X–Y stage in both axial directions. The above derivations exclude the effects of resistor R_1, which is commonly included to prevent the drifting effects of the capacitance after prolonged operations.

The capacitance of MEMS-based devices is usually in the pico or even femto Farad region. When actuated in micro or even nanometer, the signal-to-noise ratio (SNR) will be very low. As such, the modulator and demodulator in Figure 12.9 are added to reduce the sensor noise level. An external sinusoidal modulator signal is essential to achieve high SNR capacitive sensing and to actuate the MEMS micro X–Y stage for obtaining position information by artificially vibrating the MEMS micro X–Y stage at a frequency ω for capacitive self-sensing during calibration of the proposed CSSA. It should be noted that a sinusoid of frequency higher than the closed-loop bandwidth of the MEMS micro X–Y stage should be used during operations (after calibration) to prevent the vibration and excitation of the stage.

12.3.2 Experimental Verification

For experiments, capacitors in the pico Farad range as well as a sinusoidal modulator signal of 5 kHz are used. The demodulator, LPF of corner frequency at 10 kHz, and the interpolator in (12.11) are implemented on a dSPACE digital control implementation system at a sampling frequency of 400 kHz. The experimental results of proposed CSSA showing the linearity of output DC (direct current) voltage v_{DC} and change in capacitance ΔC are shown in Figure 12.10 [13] and an identified linear relationship of is obtained.

$$v_{DC} = -0.07\Delta C - 0.11. \tag{12.12}$$

It can be seen from Figure 12.10 that the output DC voltage is linear with change in capacitance which is, in turn, proportional to the displacement of the MEMS micro X–Y stage. The proposed CSSA scheme calibrated with the measurements from the high-speed cameras is used for absolute displacement sensing in both axial X and Y directions.

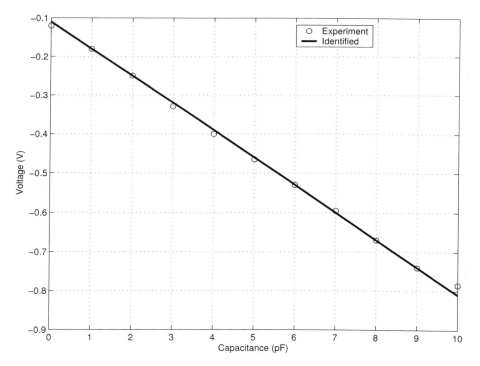

FIGURE 12.10
Experimental results of proposed CSSA. (From Ref. [10].)

MEMS Micro X–Y Stage Media Platform

12.4 Robust Decoupling Controller Design

With the displacement sensing issue for the axial X- and Y-axes resolved, a robust decoupling controller design methodology is presented in this section. The designed digital multi-input-multi-output (MIMO) controller is evaluated with simulation results.

12.4.1 Choice of Pre-Shaping Filters

To ensure robustness against parametric uncertainties in the actual micro X–Y stage during fabrication process and release, the "clean" high frequency double integrator properties of MEMS actuators are exploited. The H_∞ loop shaping method for MIMO controller design detailed in [9] will be modified and explored here. This one step approach used as the open-loop bandwidth (gain crossover frequency) is commonly used as a measurement of error rejection capabilities in many precision motion control systems such as data storage systems and robotic systems. Moreover, the loop shape at a gain crossover frequency f_c of 4.5 kHz (about ten times above the first in-plane sway mode of the micro X–Y stage at 440 Hz) can be improved for better gain and phase margins.

Before the robust stabilization of the micro X–Y stage with normalized left coprime factorization is carried out, shaping filters $W_1(s)$ and $W_2(s)$ are cascaded before and after the micro X–Y stage $G(s)$ to shape the largest singular values into the shaped plant $G_s(s)$ as

$$G_s(s) = W_2(s)G(s)W_1(s). \tag{12.13}$$

For our application, $W_1(s)$ is chosen as an identity matrix to ease controller synthesis and implementation. $W_2(s)$ is chosen as a notch filter to compensate for the small damping ratio and large gain at the resonant frequency of 440 Hz, and a lead compensator at f_c for closed-loop stability. A decoupler is also included to further reduce the axial coupling for independent axial control. The shaping filter also prevents any numerical stability while synthesizing the robust controller and is obtained as

$$W_2(s) = \frac{s^2 + 2 \cdot 0.0009 \cdot (2\pi 440)s + (2\pi 440)^2}{s^2 + 2 \cdot 1 \cdot (2\pi 440)s + (2\pi 440)^2} \times K_c \cdot \frac{s + \dfrac{2\pi f_c}{2\alpha}}{s + 2\pi}$$

$$\cdot \frac{s + \dfrac{2\pi f_c}{\alpha}}{s + 2\alpha 2\pi f_c} \cdot \begin{bmatrix} 1 & -0.006 \\ -0.006 & 1 \end{bmatrix} \tag{12.14}$$

where $5 < \alpha < 10$ is used typically [8]. For the simulation, we have chosen $\alpha = 6$ and by setting $|W_2(j2\pi f_c)G_{xx}(j2\pi f_c)| = 1$, K_c is tuned to ensure maximum phase lead at desired f_c. The frequency response of the shaping filter $W_2(s)$ is shown in Figure 12.11.

12.4.2 Controller Synthesis

Using the designed shaping filter $W_2(s)$, the robust stabilization controller $K(s)$ is synthesized by solving the robust stabilization problem, assuming that the shaped plant $G_s(s)$ has a normalized coprime factorization of $G_s = M_s^{-1}N_s$. It should be noted that the Laplace variable s has been omitted for simplicity in notation but without loss of generality.

Assume that the micro X–Y stage is perfectly decoupled by the decoupler in $W_2(s)$. The controller design can now be carried out independently for each axis with the diagonal and symmetrically shaped plant $G_s(s)$. With A, B, C, and D as the system's dynamics state-space matrix quadruple of the micro X–Y stage's identified mathematical model shown previously in (12.2),

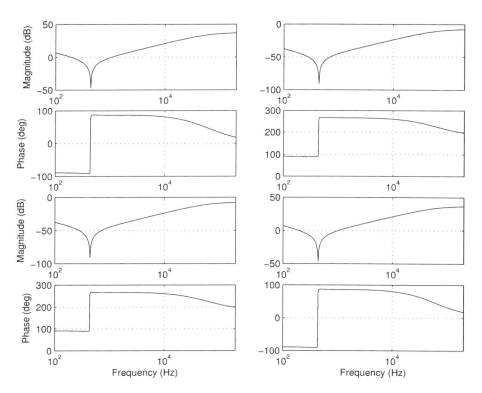

FIGURE 12.11
Frequency responses of $W_2(s)$. (From Ref. [10].)

MEMS Micro X–Y Stage Media Platform

181

the micro X–Y stage can be represented by a strictly proper system with no feedthrough, i.e., $D = 0$. As such, the controller K_x which ensures that

$$\left\| \begin{bmatrix} K_x \\ I \end{bmatrix} (I - G_{xs}K_x)^{-1} M_{xs}^{-1} \right\|_{\infty} \leq \gamma \tag{12.15}$$

for a specified $\gamma > \gamma^*$ (γ^* is the H_∞ norm of (12.15)), is given by [9,14]

$$K_x = \begin{bmatrix} A + BF + \gamma^2 \left(L^T\right)^{-1} ZC^T C & \gamma^2 (L^T)^{-1} ZC^T \\ B^T X & 0 \end{bmatrix} \tag{12.16}$$

where G_{xs} is the shaped micro X–Y stage in the X-axis, M_{xs} is the left coprime factorization of G_{xs}, and I is the identity matrix of appropriate dimensions. F and L are given by

$$F = -B^T X \tag{12.17}$$

$$L = \left(1 - \gamma^2\right)I + XZ \tag{12.18}$$

with Z and X being the unique positive definite solutions to the following Algebraic Riccati Equations (AREs), respectively.

$$AZ + ZA^T - ZC^T CZ + BB^T = 0 \tag{12.19}$$

$$A^T X + XA - XBB^T X + C^T C = 0. \tag{12.20}$$

Obviously, Z and X can also be obtained by performing Schur's complements and solving with linear matrix inequalities techniques. On controller synthesis, a $\gamma^* = 1.4705$ is obtained. By choosing a $\gamma > \gamma^* = 2.0$ which results in suboptimal H_∞ controller, a 50% perturbation of magnitude of coprime uncertainty can be tolerated before instability. The frequency responses of the synthesized $K_x(s)$ are shown in Figure 12.12.

The synthesized $K_x(s)$ effectively reduces the low frequency gain and the gain crossover frequency slightly, while increasing the roll-off at high frequencies for robust stability. The final controller $K(s)$ for the micro X–Y stage is given as follows:

$$K(s) = \begin{bmatrix} K_x(s) & 0 \\ 0 & K_y(s) \end{bmatrix} \cdot W_2(s) \tag{12.21}$$

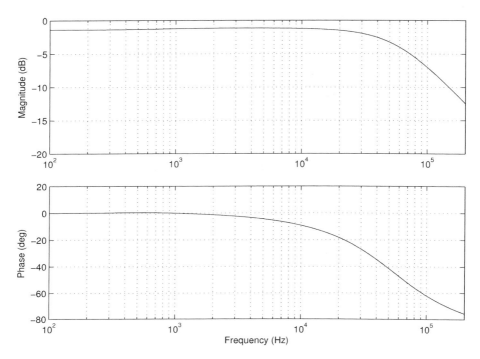

FIGURE 12.12
Frequency responses of $K_x(s)$. (From Ref. [10].)

and $K_y(s) = K_x(s)$ after the decoupling control. The frequency response of $K(s)$ is shown in Figure 12.13.

12.4.3 Frequency Responses

The frequency responses of largest singular values in shaped plant $G_s(s)$ and open-loop transfer function $K(s)G_s(s)$ are plotted in Figure 12.14. It can be seen that the overall loop shape is generally unaltered, but a gentler slope is observed near the crossover frequency f_c translating into larger stability margins.

The block diagram for digital control of micro X–Y stage in both axes is shown in Figure 12.15. The two-input-two-output control $K(z)$ is obtained by discretizing $K(s)$ at sampling frequency of 400 kHz with bilinear transformation.

12.4.4 Time Responses

For step responses, the overshoot of the micro X–Y stage should be kept to a minimum. As such, a nonlinear saturator is added to $K(z)$ to ensure that the

MEMS Micro X–Y Stage Media Platform

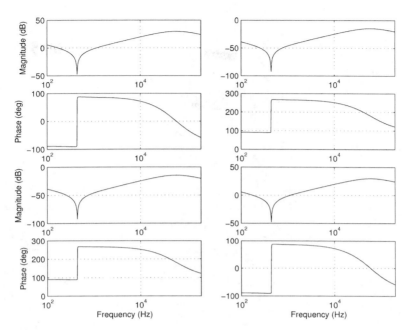

FIGURE 12.13
Frequency responses of $K(s)$. (From Ref. [10].)

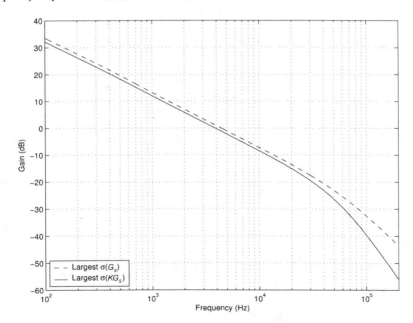

FIGURE 12.14
Plots of largest singular values. Dashed: $G_s(s)$. Solid: $K(s)G_s(s)$. (From Ref. [10].)

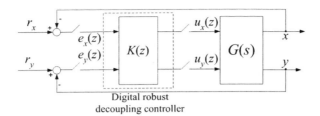

FIGURE 12.15
Block diagram for digital control of micro X–Y stage. (From Ref. [10].)

control signal does not exceed the maximum allowable voltage of 55 V. $K(z)$ must also be able to decouple the micro X–Y stage such that the interaction in the main axes is kept to a minimum. The step response of the micro X–Y stage with 20 μm for 2 ms followed by 14 μm in the X-axis (or r_x) is shown in Figure 12.16. The step response of the micro X–Y stage with 14 μm at 1 ms for 2 ms followed by 6 μm in the Y-axis (or r_y) is shown in the same figure to observe the mechanical crosstalk of the micro X–Y stage. It can be seen that

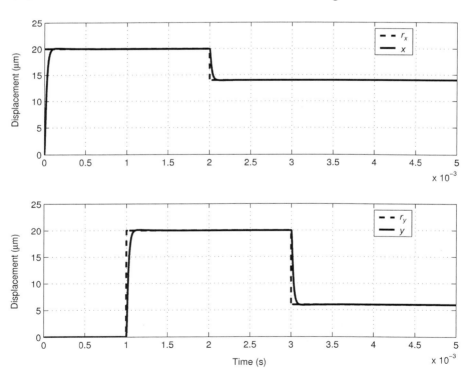

FIGURE 12.16
Simulated step responses. (From Ref. [10].)

the actuation in both axes is decoupled and the interactions in between both axes are in the orders of magnitudes 10^{-12} m.

The corresponding control signals are shown in Figure 12.17. It can be seen that the large span seek operations with little overshoot are achieved within the control signal limitations.

12.4.5 Robustness Analysis

Most MEMS actuators demonstrate double integrator properties at high frequencies with little uncertainties. As such, only the largest singular value of $G(s)$ is perturbed with ±50% to demonstrate the robustness of the synthesized digital controller $K(z)$. The step responses are simulated with the reference sequences r_x and r_y again and shown in Figure 12.18. It can be seen that the closed-loop digital control system is robustly stable with the controller $K(z)$ to ±50% change in the largest singular value of the MEMS micro X–Y stage.

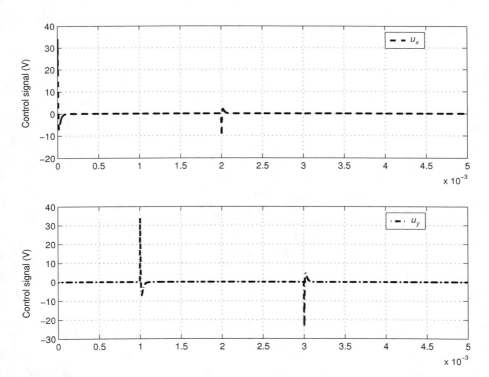

FIGURE 12.17
Control signals. (From Ref. [10].)

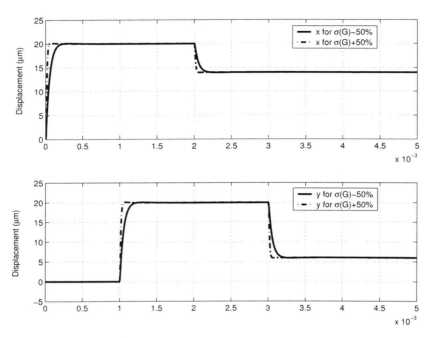

FIGURE 12.18
Simulated step responses for perturbed system. (From Ref. [10].)

12.5 Conclusion

In this chapter, the MEMS micro X–Y stage with 6 mm × 6 mm recording media platform has been introduced and proposed to integrate 40 nm PMMA film recording media. The displacement of the media platform is 20 μm with the driving voltage of 55 V. A CSSA scheme has been proposed for sensor fusion and verified with experimental results. With the first two resonant in-plane modes of the MEMS micro X–Y stage at 440 Hz, a robust decoupling control scheme has been proposed. The simulation results have shown that the digital servo MIMO system has strong error and vibration rejection capabilities.

References

1. E. Eletheriou, T. Antonakopoulos, G. K. Binning, G. Cherubini, M. Despont, A. Dholakia, U. Durig, M. A. Lantz, H. Pozidis, H. E. Rothuizen, and P. Vettiger, Millipede - A MEMS-based scanning-probe data-storage system, *IEEE Trans. Magn.*, 39(2), pp. 938–945, March 2003.

MEMS Micro X–Y Stage Media Platform

2. P. Vettiger, M. Despont, U. Drechsler, U. Durig, W. Haberle, M. I. Lutwyche, H. E. Tothuizen, R. Stutz, R. Widmer, and G. K. Binnig, The "Millipede"- More than one thousand tips for future AFM data storage, *IBM J. Res. Dev.*, 44(3), pp. 323–340, May 2000.

3. P. Vettiger and G. K. Binnig, The nanodrive project, *Sci. Am.*, 228(1), pp. 46–53, January 2003.

4. A. Pantazi, M. A. Lantz, G. Cherubini, H. Pozidis, and E. Eleftheriou, A servo-mechanism for a micro-electro-mechanical-system based scanning-probe data storage device, *Nanotechnol.*, 15(10), pp. S612–S621, October 2004.

5. J. Choi, H. Park, K. Y. Kim, and J. U. Jeon, Electromagnetic micro X–Y stage for probe-based data storage, *J. Semicond. Technol. Sci.*, 1(1), pp. 84–93, March 2001.

6. C. Kim and Y.-K. Kim, Micro XY-stage using silicon on a glass substrate, *J. Micromech. Microeng.*, 12(2), pp. 103–107, March 2002.

7. J. F. Alfaro and G. Fedder, Actuation for probe-based mass data storage, *Technical Proceedings of the Fifth International Conference on Modeling and Simulation of Microsystems*, San Juan, Puerto Rico, April 22–25, 2002, pp. 202–205.

8. C. K. Pang, G. Guo, B. M. Chen, and T. H. Lee, Self-sensing actuation for nanopositioning and active-mode damping in dual-stage HDDs, *IEEE/ASME Trans. Mech.*, 11(3), pp. 328–338, June 2006.

9. K. Glover and D. McFarlane, Robust stabilization of normalized coprime factor plant descriptions with H_∞ bounded uncertainty, *IEEE Trans. Autom. Control*, 34(8), pp. 821–830, August 1989.

10. C. K. Pang, Y. Lu, C. Li, J. Chen, H. Zhu, J. Yang, J. Mou, G. Guo, B. M. Chen, and T. H. Lee, Design, fabrication, sensor fusion, and control of a micro X–Y stage media platform for probe-based storage systems, *Mech.*, 19, pp. 1158–1168, 2009.

11. R. Legtenberg, A. W. Groeneveld, and M. Elwenspoek, Comb drive actuators for large displacement, *J. Micromech. Microeng.*, 6(3), pp. 320–329, September 1996.

12. M. Xie, X. Wang, M. Yu, M. Zhang, and G. Wang, Analysis of the air damping in MEMS lateral driven microresonators, *Proc. of the 9th International IEEE CPMT Symposium on High Density Design, Packaging and Microsystem Integration (HDP'07)*, Shanghai, P. R. China, June 26–28, 2007, pp. 1–4.

13. Li C, Design and implementation of position error signals for probe-based storage systems. BE thesis, National University of Singapore, 2005.

14. K. Zhou and J. C. Doyle, *Essentials of Robust Control*, Prentice Hall, Upper Saddle River, NJ, USA, September 1997.

13

Conclusions

Many secondary actuators have been developed in addition to primary actuators in the field of mechanical actuation systems. The aim is to provide high performance such as high precision and fast response. Several types of secondary actuators have been introduced such as PZT (Pb-Zr-Ti) milliactuator, electrostatic microactuator, PZT microactuator, and thermal microactuator. Comparison of these secondary actuators has been made, and these secondary actuators have made dual- and multi-stage actuation mechanisms possible.

Three-stage actuation systems have been proposed for the demand of wider bandwidth, to overcome the limitation by stroke constraint and saturation of secondary actuators. After the characteristics of the three-stage systems have been developed and the models have been identified, the control strategy and algorithm have been developed to deal with vibrations and meet different requirements. Particularly, for the three-stage actuation systems, the presented control strategies make it easy to further push the bandwidth and meet the performance requirement. The control of the thermal microactuator based dual-stage system has been discussed in detail, including linearization and controller design method.

The developed advanced algorithms applied in the multi-stage systems include H_∞ loop shaping, anti-windup compensation, H_2 control method, and mixed H_2/H_∞ control method. Typical problems of the milli- and microactuators as the secondary actuators have been considered and appropriate solutions have been presented such as saturation compensation, hysteresis modeling and compensation, stroke limitation, and PZT self-sensing scheme. Time delay and sampling rate effect on the control performance have been analyzed to help select appropriate sampling rate and design suitable controllers.

Specific usage of PZT elements has been produced for system performance improvement. Using PZT elements as a sensor to deal with high-frequency vibration beyond the bandwidth has been proposed and systematic controller design methods have been developed. As a more advanced concept, PZT elements being used as actuator and sensor simultaneously has also been addressed in this book with detailed scheme and controller design methodology for effective utilization.

Enabled by MEMS (microelectromechanical systems) microactuators, the micro X–Y stage has been developed for nanopositioning, and the fabrication process has been detailed. Both static and dynamics modeling of the micro X–Y stage have been presented. On the basis of the designed feedback control, specific problems such as crosstalk decoupling have been addressed and solved.

Throughout the book, simulations and experiments with dual-stage, three-stage systems, and micro X–Y stage platform were performed and detailed evaluation results were presented. Extremely high bandwidth up to 10 kHz has been achieved under different levels of constraint coming from the vibration and system itself. Therefore, this book has provided control design methodologies to solve various problems and meet performance requirements for multi-stage actuation systems working under practical conditions. Using the methods in this book, positioning accuracy will be further improved for the next generation of high-performance motion control systems required for advanced manufacturing, cyber-physical systems, and internet-of-things (IoT) platforms.

Index

A

Active vibration control, 57, 134
Air-bearing surface, 2
Air damping, 171–172
Anti-windup saturation compensator
    description of, 79
    design of, 80–82
    simulation and experimental results,
        82–87

B

Bode's integral theorem, 133
Bridge circuit, 155
    design of capacitive self-sensing
        actuation, 176–177
    self-sensing actuation and, 156–157
Buffered oxide etching (BOE), 174

C

Capacitive self-sensing actuation
    (CSSA), 167, 168, 176–178, 186
Closed loop control system
    hysteresis compensation, 114, 116
    sensitivity function of, 15, 16, 18
    servo bandwidth, 134
    stages of, 60
    three-stage system, experiment
        results, 60–61
    vs. with/without hysteresis
        compensation, 116
Complementary metal oxide
    semiconductor (CMOS), 173
Controller design method, 141–143
    continuous-time domain, 18–22
    controller synthesis, 180–182
    and control strategy, 52–56
    in discrete-time domain, 22–25
    fast controller and subsystem
        estimator, 137
    frequency responses, 182

mixed H2/H-infinity(∞) controller
    cd(z)
    modeling with time delay for, 91–95
    notch filters and h_2 method, 143–144
    pre-shaping filters choices, 179–180
    PZT controller, 161–163
    robust decoupling, 179–186
    slow controller, 137–138
    time responses, 182–185
    voice coil motor controller, 160–161
Control loop
    active damping, 141
    bandwidth of, 4
    design, 99
    dual-stage system, 166
    feedback, block diagram of, 142
    against microactuator saturation, 39
    servo, 50, 91–92
    voice coil motor, 54
Control scheme, 138, 139
    block diagram of, 142, 145
    decoupled, 33, 52
    for dual-stage actuation system,
        15–18
    robust decoupling, 167, 168, 186
Control system
    closed-loop, 8, 45, 61, 185
    configurations of, 61–66
    dual-stage
        bandwidth of, 76
        requirement of, 26
        sensitivity function of, 28, 30, 37,
        41
        step response of, 30
        structure of, 121–123
    overall design, 52
    real-time digital, 89
    sampling rate for, 25, 54
    servo delay in servo, 91
    singular perturbation-based, block
        diagram of, 135
CSSA, *see* Capacitive self-sensing
    actuation (CSSA)

191

Index

D

Damping control method, 143–146
Decoupled control scheme, 33, 166
Decoupled control structure, 16, 45, 52
Decoupled master–slave structure, 37, 87
Deep reactive ion etching (DRIE), 173
Digital control implementation system (dSPACE), 110, 113, 150, 178
Digital control system, 89, 137
  robustness analysis, 185, 186
Digital notch filters, 25, 133, 143–144, 147, 166
Discrete-time domain, controller design method in, 22–25, 143
Dual-stage actuation systems, 7–8
  block diagram of, 9
  controller design of, 25–29
  control scheme for, 15–16
  description of, 2
  mechanism, 133
  parallel control structure for, 18
  PZT elements in, 150
  saturation control for microactuators in, 79–87
  secondary actuator of, 69–76
  seeking
    controls of, 119–131
    performance of, 43
  sensitivity function of, 28, 30
  servo control loop, block diagram of, 92
  thermal microactuator using, 33–45
  time delay and sampling rate effect on, 89–103
  voice coil motor actuator of, 25

E

Electronics card, 2
External vibration, 60, 65
  actuators' control effort, 57
  and microactuator stroke, 74–76
  spectrum of, 50, 51
  stroke and, levels, 58
  system working environment, 57

F

Fast Fourier Transform (FFT), 114, 116
Feedback control loop, 133, 141
Finite element analysis (FEA), 167
Fminsearch (Matlab function), 124

G

Generalized Prandtl-Ishlinskii (GPI), hysteresis
  mathematical approach, 117
  modeling, 105–109
  PZT-actuated structure, 109–116

H

H2/H-infinity(∞) method, 134, 144–146, 149, 150, 152
Hysteresis, PZT-actuated structure
  compensator design, 112
  experimental verification, 112–116
  modeling of, 105–109, 110–112

L

Laser Doppler vibrometer (LDV), 25, 29, 34, 41, 60, 84, 114, 136, 159
Linear matrix inequality (LMI)
  approach, 20–21, 23, 26, 39, 79, 81–82, 134, 135, 146
Linear quadratic Gaussian (LQG), 134
Loop shaping method, discrete-time H-infinity(∞), 22, 29, 31, 38–40, 53–54, 76, 179

M

Mechanical actuation systems
  dual-stage, 2, 9
  micro X–Y stage, 3–4
  overview of, 1
  primary actuators, 4–6
  secondary actuators, 6–8
  single-stage, 8–9
  three-stages of, 10–11

Index

MEMS, *see* Microelectromechanical systems (MEMS)

Microactuator(s)
    controller design, freedom loop shaping for, 70
    control scheme, third stage, 52
    loop, 10, 37, 42, 54, 65, 70, 75, 76
    microelectromechanical systems, 189–190
    saturation control for, 79–87
    secondary actuators, 7–8
    stroke, 57, 58, 74
    suspension assembly, 33
    thermal (*see* Thermal microactuator)
    third-stage, 52, 54, 63, 65

Microelectromechanical systems (MEMS), 3, 167–168, 189
    capacitive self-sensing actuation, 176–178
    micro X–Y stage
        design and simulation of, 168–171
        fabrication of, 173–175
        modeling of, 171–173
    robust decoupling controller design, 179–186

Microelectromechanical systems (MEMS) actuator, 87

Micro thermal actuator (MTA), 34, 35, 38, 41, 43, 44

Micro X-Y stage, microelectromechanical systems
    block diagram, digital control of, 182, 184
    capacitive self-sensing actuation, 176–178
    design and simulation of, 168–169
    dynamic performance of, 170–171
    fabrication of, 173–175
    mechanical actuation systems, 3–4
    modeling of, 171–173
    novel design of, 167
    robust decoupling controller design, 179–186
    static analysis of, 169–170

MTA, *see* Micro thermal actuator (MTA)

Multi-input-single-output (MISO), 21

N

"Nanodrive," 3, 167, 168

Nanometer-wide tips, of probes performance, 4

Nonlinear least-square optimization, 111, 112

Nonlinear loop, 80, 81

O

Open-loop hysteresis compensation structure, 112–114

Open-loop transfer functions
    dual-stage, 17, 29, 53, 70
    frequency responses of, 26, 54, 64, 94
    three-stage actuation system, 53, 55
    voice coil motor actuator, 40, 93

Oxygen plasma, 173

P

Pade approximation, 90, 91

PBSS, *see* Probe-based storage systems (PBSS)

Pb-Zr-Ti (PZT)
    actuated structure, 109–116
    controller, 27, 161–163
    displacement estimation circuit H_B, 157–160
    for dual-stage actuation decoupling, 70
    fast dynamics identification, sensor, 135–136
    hysteresis
        actuated structure, 109–116
        modeling, 105–109
    milliactuator, 6, 47, 57, 121, 189
        controller of second-stage, 48, 54
        frequency responses of, 71, 93
        hard disk drive with, 2
        loop, 58, 59, 76, 93
        seeking within, stroke, 127–131
        open-loop $Gp(z)$ frequency responses, 72, 73, 93
        secondary actuators, 6–7, 25, 156, 163–165, 189

Pb-Zr-Ti (PZT) (*cont.*)
secondary actuator's displacement, online estimation of, 156–160
sensor, 134, 147, 148
vibration control, damping control method, 133–134
application results, 146–150
controller design, 141–143
singular perturbation method, 134–141
Polymethyl methacrylate (PMMA), 167, 173–174, 186
Position error signals (PESs), 2, 91
Positioning information, 2
Positive position feedback (PPF), 161
Preisach model hysteresis, 106
Primary actuator, voice coil motor, 4–6, 21, 31, 66, 119–121, 160
Probe-based storage systems (PBSS), 3, 167
Proportional-integral-derivative (PID), 25, 29, 37, 137
Proximate-time-optimal-servomechanism (PTOS) method, 119, 120

R

Robust decoupling controller design, 179–186
Robust decoupling control scheme, 167, 168, 186
R/W/E operations, 4

S

Sampling rate effect, dual-stage actuation systems, 99–103
Saturation control, for microactuators in dual-stage actuation systems, 79–87
Scanning electron microscopy (SEM), 174
Secondary actuators
microactuators, 7–8
PZT milliactuator, 6–7, 156, 189
stroke constraint/saturation of, 47, 189
Seeking control structure, block diagram of, 122

Self-sensing actuation (SSA)
design of controllers, 160–163
dual-stage control topology, 155
performance evaluation, 163–166
PZT secondary actuator's displacement, online estimation of, 156–160
Sensor signal measurement delay, 95, 96, 98, 99, 103
Servo bandwidth, 1, 9, 10, 79, 95, 101, 134
Servo control performance, 97–99
Signal conditioning amplifier, 135
Signal- to-noise ratio (SNR), 155, 177
Silicon-on-insulator (SOI), 169
Single-stage actuation system, block diagram of, 8
Singular perturbation theory, 135–141, 152
SSA, *see* Self-sensing actuation (SSA)

T

Thermal microactuator, dual-stage actuation system
controller design/performance evaluation, 37–41
experimental results, 41–44
modeling of, 33–36, 79–80
stroke, 66
Three-stage actuation system, 189
actuator and vibration modeling, 47–52
block diagram of, 10, 11
control
strategy and controller design, 52–56
system configurations, 61–66
evaluation performance, 56–60
experiment, closed-loop control of, 60–61
Time delay, dual-stage actuation systems
for controller design, 91–95
effects of, 95–99
modeling of, 90–91
and sampling rate, 99–103, 189
Trajectory optimization, 119, 122
control system structure, 121–123

Index

voice coil motor current profile and dual-stage reference, 123–130

Two degrees of freedom (2DOF), 119

U

Uncertainty, 143, 147, 149

V

Vibration control, damping control method
  application results, 146–150
  controller design, 141–146
  singular perturbation method, 134–141

Vibration modeling, actuator and, 47–52

Voice coil motor (VCM), 156
  actuator
    frequency responses of, 25, 34, 48, 54, 92
    minimum-jerk current profiles of, 121, 123, 129
  controller, 27, 37, 160–161

dual-stage actuation systems, 25
loop and secondary actuator loop, 16, 17
plant, frequency responses of, 36
primary actuator, 21, 160
  current profile, 120–121, 123–127
resonance modes, 134
three-stage actuation system, sensitivity functions of, 55
vibration control, damping control method, 133–134
  application results, 146–150
  controller design, 141–146
  singular perturbation method, 134–141

W

Weight(ing) function, 15, 19–22, 24, 38, 39, 70–72

Z

Zero-order-hold (ZOH), 92
"Zero-order-hold" method, 25–26